I0504819

Light Years Away

Light Years Away

✦

The Whole Creation at a Glance

Sudhir K. Routray

iUniverse, Inc.

New York Lincoln Shanghai

Light Years Away
The Whole Creation at a Glance

All Rights Reserved © 2004 by Sudhir Kumar Routray

No part of this book may be reproduced or transmitted in any form or by any means, graphic, electronic, or mechanical, including photocopying, recording, taping, or by any information storage retrieval system, without the written permission of the publisher.

iUniverse, Inc.

For information address:
iUniverse, Inc.
2021 Pine Lake Road, Suite 100
Lincoln, NE 68512
www.iuniverse.com

ISBN: 0-595-33582-9

Printed in the United States of America

To My Mother

Contents

Preface

The idea of writing this book came when I was going through an awesome article on a supernova and its radiation characteristics. Before that I had read a lot of articles, magazines like the Scientific American and books on space and space exploration. I had found enough on specific areas in those books and magazines. But there was not a book which has everything from the sun to the remote end of the universe and beyond that. After reading about the planets, stars, galaxies, various models of the universe and the possibilities beyond our universe I thought it would be a better idea to put all these things in a book, which would cover the whole things (of course not everything as it is not possible to cover everything in such a small book). Sometimes when I am free I think about the possibilities of the other universes beyond ours. From those thoughts I have embedded some of them in this book. Especially the last chapter is exclusively full of those thoughts. Overall this book has been written for the enthusiasts of space, Astronomy and Physics.

Anyway, thanks a lot for reading this preface. Hope you will enjoy reading this book. It is quite informative for the students and those who do not want to go through the complexities of mathematics. Intentionally, the mathematical things, complex theories and calculations have been avoided in this book. But there are enough references for those who want to go through every detail of the available things. Similarly photos of various cosmic bodies have been avoided because there are a lot of photos (more than a million) and it is not possible to present all of them here. Another reason is that all those photos are not enough to understand the universe rather a physical insight is more important. Alternatively, the sources of photos and archives have been given in the references.

Introduction

Light is a form of energy which gives us the perception of vision. In the early age people were worshiping the sources of light. Light was thought to be a source of knowledge and power. There were also many myths about the speed of light. For the first time it was predicted by Galileo to be travelling at a very high speed. Galileo himself had tried many procedures to measure the speed of light. The most famous of them was, using two lanterns separated by some miles. Then many people tried many other methods to measure the velocity of light. But the success came quite late in the early 20th century. The Nobel winning experiment of Michelson and Morley proved that the speed of light is approximately 3×10^8 m/s. Before that the theoretical foundation was given by Sir J. C. Maxwell. Maxwell was a great theorist of the electrical sciences. He wrote his historic equation on the speed of light in 1869. He proved that light is a kind of electromagnetic wave and it obeys all the equations that are obeyed by the other electromagnetic waves. He also unified the optics as a branch of electromagnetism as a whole. The milestones on the speed of light (in brief) have been given below.

- In 1667 Galileo tried to measure the speed of light and said it is approximately 10 times faster than sound

- In 1675 Ole Roemer calculated the speed of light to be 200 000 km/s. It was based on the calculations of the eclipses of Jupiter's moons

- 1728 James Bradley calculated the speed of light to be 301 000 km/s

- In 1869 Maxwell theoretically calculated the speed of light to be $1/\sqrt{(\epsilon\mu)}$

- In 1849 Fizeau calculated the speed of light to be 313 000 km/s

- In 1887 Michelson and Morley through their historic experiment found the speed of light is almost 186,355 miles/sec
- In 1926 Foucault found the speed of light to be 229 796 km/s. His experiment was quite similar to that of Fizeau's
- Quantum calculations confirm the speed of light to be 299792.49 km/s. But according to Feynman more than one speed of light is possible depending on the way the photons move
- Today it is **299792.5 Km/s** (the most used value) with some factors of correction

Since those days of Michelson and Morley, the speed of light has always attracted the physicists or other scientists for some reason or the other. Looking at the enormous value of the speed of light many astronomers and physicists tried to make the speed of light as a unit of distance. That means the distance travelled by light in some time interval can be regarded as a unit of distance. That was an amazing idea in the mind of some researchers. Along with the light year, the other units, which are used as the unit of measurement of the astronomical distances, are AU (astronomical unit) and Parsec. One AU is the average distance between the sun and earth. Similarly, one Parsec is one parallax second. The details of parallax and its importance in astronomy have been discussed in <u>appendix-A</u>. In terms of basic units of length, one light year can be calculated in the following easy steps.

1 light year = 365 × 24× 60 × 60 × speed of light = 9.4607×10^{15} metres (approximately)
1 light-year = 63240 AU (approximately)
1 light-year is about 9.4607×10^{15} metres and
1 astronomical unit (AU) is equivalent to the average radius of the Earth's orbit
So, 1 AU is about 1.4960×10^{11} m = $1.58128588 \times 10^{-05}$ light years
1 Parsec = $3.08568025 \times 10^{16}$ meters = 3.26163626 light years (more can be found in appendix-D)

In general (according to a layman's common sense), light means, we understand the visible light, which helps us to see our surroundings and the outside world. In fact visible light is just a small fraction of the whole light spectra (electromagnetic radiation). In terms of energy also visible light is very weak in comparison to its stronger cousins like X-ray and Gamma ray. Light has many special characteristics, which makes its study so wonderful and fascinating. These special properties of light have been presented in the <u>appendix-B</u>. Whatever we know about the outer space, is only possible due to the light that we get from those remote objects.

In the subsequent chapters the mysterious sciences of the things (cosmos and the universe as a whole) which are light years away from us have been presented. It has been started from our sun till the remote end of the universe and beyond that. Light really provides us a great means to know all these facts. Various other issues of the universe and its structure like the inflation theory, big bang and other common cosmic phenomena have been looked at. In all these facts, some of the ideas and possibilities have been embedded in a scientific way. Especially in the last two chapters the possibilities of the extra-universal thing have been presented, from where no light can come to us and, the only means of studying those areas is by imagination or human thoughts.

1

Our Sun and Its Surrounding

Since times immemorial, we have been seeing many things in the sky without any addition to our eyes. What we see, are nothing but the nearest cosmic objects to our earth. Some of them are quite luminous and some are not. Due to the luminosity some of the distant objects catch our attention but some nearer bodies go ignored. But the objects what we see since ages are sun, moon and some of the brightest stars and planets.

A little before 4.6 billion years ago there was no sun, no moon, no earth no solar system or nothing what we see around us. A whirling nebula started its fate to be some future stars. Our sun was one of those whirling bodies that went through the long evolution to come to the present state. The dust and moving masses started gathering around some massive centres (due to gravity). When the amount of the gathered mass increased, there was an increasing pressure at the centre which increased the temperature to a very high value. The gravity was also increasing in a proportionate rate. When the temperature became high enough to make the electrons of the atoms to be free from their nuclei the random motion increased and more heat was generated. At some time, the temperature becomes high enough to start the fusion process. The scenario was changed, and our sun and other stars around it started producing a huge amount of energy. The energy was in various forms like heat and light. It made the stars hot, bright and energetic. The smaller parts which were not able to start the nuclear ignition process remained lack lustre through out their life. They were either the planets or brown dwarfs. The smaller bodies were comets, asteroids or more chaotic stellar bodies like the objects of the Kuiper belt.

More or less the formation of planets was around the same time as our sun. But some theories say that the planets were formed long after the formation of the sun (in the process of rotation or collision, perhaps planetary bodies were disintegrated from the stars). It may be true. But anyway, it happened a long time ago. Looking at the distance of various planets from the sun it seems that the planets and the sun were the result of the same process which had started almost 4.6 billion years ago. But it took a long time for the solar system to be like what we see today. The cooling process was very slow and it allowed various states to be settled down.

As we see today the sun and its family, the solar system, is the result of that prolonged past. What really happened in the whole time scale is really difficult to say as there are many theories to explain this fact and they differ from one another in many ways. All these theories can be divided into the two following broad categories.

1.   Catastrophic theories and

2.   Non-catastrophic theories

According to the catastrophic theories, as the name indicates, the solar system was originated from some kind of catastrophe. Different scientists and astronomers who believe the catastrophic origin suggest various kinds of catastrophes, which created the solar system. According to these theories sun and other members of the solar system were not created at the same time rather the sun was the first and oldest member of the solar system and the planets came after the catastrophes. The most possible catastrophes are the collisions (or when a star come close to the sun, it has a tendency to pull out matter from sun) with stars and other bigger heavenly bodies. On the other hand, the non-catastrophic theories predict that the sun and planets were formed from the same physical process and the only difference came when the mass was different for different stellar bodies. Sun was too massive to be a star while the planets were light enough to remain as they are today (of course they have under gone a lot of changes since their origin). Among the non-catastrophic theories the solar nebula theory is the most popular theory now.

Anyway here we are not going through the details of all those theories behind the origin of the solar system. Because those debates are too long to be presented here. Only brief ideas have been given above. In the next section we are going through the details of the solar system, its members, their behaviour and their interactions.

Now first of all, let us have a look at the solar system and its members. The solar system can be divided into two parts according to the distance and state of its planets. The inner part is known as, 1) Inner solar system and the outer part is, 2) Outer solar system. In the inner solar system the planets are Mercury, Venus, Earth and Mars. Similarly, the rest five planets are in the outer solar system. In terms of order and discipline the inner solar system is very stable and predictable. Chaos is minimum in this part of the solar system. But after Mars the scenario is different. The asteroid belt starts after Mars and the number of asteroids is maximum in between Mars and Jupiter. The behaviour of the objects and stellar bodies is highly chaotic in this region and beyond this. The first four planets of the inner solar system are also quite different from the view point of their formation and state. They are small and rocky but the outer solar system planets are large and gaseous giants (except Pluto). There are many reasons behind all these diversities. We shall go through all of them in their individual studies in the following sections.

Sun

Let us start the story of our solar system with our sun itself. It is the heaviest object of the solar system having about the 99.8% of the whole mass of the system. In terms of distance it is 8 light minutes from us. That means the solar radiations that we get here on earth, actually left sun before 8 minutes. This distance is also known as AU or Astronomical Unit. In terms of meter it is a big number written below. This is actually the mean distance of our earth from the sun. The orbit of earth is not a circle; rather it is an ellipse with eccentricity 0.0167. We shall have a detailed look on it later in this chapter, while studying the earth.

1 Astronomical Unit = 149 598 000 000 meters

Our sun is basically classified as a G2 star (the classification of stars can be found in appendix-C). Till date it has been discovered that there are more than 100 billion G2 stars in our galaxy. So, sun is a normal star whose type is abundantly found in our galaxy. The mass of the sun is huge. It is found to be approximately 1.9892×10^{30} kilograms. Its constituents are mainly hydrogen and helium. In the process of nuclear fusion, hydrogen is constantly converted to helium. At the moment, the amount of hydrogen is almost 70% and helium is 28%. The rest 2% is various metals (an approximate figure). As the time passes hydrogen is converted into helium and helium is converted into heavier elements (but the process of helium to higher element conversion is very slow now). At the core, mainly metals like iron are there. The gravity of sun is too high to keep all these stuffs tightly bound to it.

Do not be excited, if you do not know that sun takes 25.4 days to rotate around its central axis like every active star. But the strange thing is that it is the rotation of the outer layer of the sun at its equator. At the poles the rate of rotation is different. It is up to 35 days. This strange behaviour of sun is obvious, due to its physical state and it is known as the differential rotation of sun. Sun is actually not a solid body like our earth. Let us see the differential rotation from close. Differential rotation means the different parts of a body rotate at different angular velocities. This indicates that the rotating body is not a rigid body rather a mixture of solid and fluids or only fluids or fluids and dust. Differential rotation is present in almost all galaxies (because they contain matter in all forms, a huge amount of gas and dust), proactive stars, gas planets and other gaseous cosmic bodies. The study of differential rotation is very important to understand the rearrangement in the cosmic bodies (due to conservation of angular momentum they have to rearrange within themselves). But the inner core of sun rotates as a solid body (the inner core is generally made up of heavy metals and is tightly bound due to intense gravity) and, that too at a different angular velocity. In our solar system, this type of similarities (in terms of differential rotation) can be found in the gas planets like Jupiter, Saturn and Uranus etc.

Sun's output at present is around 3.826×10^{26} watts. Sun is the largest single source of energy in the solar system. This huge amount of energy is actually generated from large scale nuclear fusion reactions though out sun. Generally it is the lower order fusion in which the hydrogen nuclei are turned into Helium nuclei. The difference in the mass between the nuclei is turned into energy according to Einstein's equation,

$E = mc^2$ (here m represents mass and c is the velocity of light)

From the above equation we can find the amount of mass converted into energy in every second. It is around 4.2×10^9 kg/sec.

The out put is in the form of radiation or photons. The very high energetic photons (like gamma ray and X-ray radiations) cannot go very far after generation (because they are produced in the bulk, not on the surface). Rather a lot of absorption (the elements like hydrogen, helium and other heavier elements absorb those radiations and get hot and thus the temperature too rises vigorously) and emission reactions (after absorbing the high energy photons they emit their corresponding low energy photons) occur and the low energy radiation in the lower frequency ranges (like ultraviolet, visible, infrared and radio) are finally emitted from the outer surface of sun. Of course a trace amount of high energetic radiations are found in the sun rays. It has been noticed that the out put of sun is not a constant; rather it varies a lot depending on the solar dynamics. The recent climate change is mainly due to the increase in the solar radiation (though the pollution and other factors are there change in solar radiation is the main reason). There are proofs of low solar radiations in the past. In the last decades of 17th century the solar radiation was minimum, for many years. At that time the winter was severe and it is also known as the little ice age, in Europe. Similarly, in the past the temperature was quite lower than what is it today.

Sun has too many mysteries. It can sallow the comets, while orbiting the sun in a very close orbit. There are many evidences of comets going behind the sun (with respect to an earth observer), while orbiting, and after that they are lost for ever. Some of the comets lose their tail when they get close to the sun. Solar storms can stop the regular functions of satellites, mobile phone networks and even the electric power networks to cause a black-out. It is due to the heavy solar winds that contain a huge number of charged particles and powerful radiations. The magnetic storms and the beautiful auroras are the effects of solar wind. The magnetic field of sun is incredibly strong and complex. Its effective magnetic zone or the magnetosphere extends beyond Pluto. The planer structure of the solar system may be due to the magnetosphere of sun.

Someday sun will also under go the same fate as the other G2 stars of Milky Way. But before coming to that end, it is predicted that sun can survive 5 billion more years. When the whole mass of hydrogen will be converted into helium and then into the heavier elements the scenario will be quite different. Sun will no more be bright as it is today; rather will under go an oscillating stage of brightness (there will be increase and decrease in the brightness of sun depending on the fuel type and consumption rate).

Sun is a star of the Milky Way galaxy and like other stars sun too revolves around the galactic centre. The speed of sun around the galactic centre is around 220km/s (the whole solar system changes the space at this speed). At this speed sun takes around 250 million years to complete one rotation. It is known as a galactic year. This is the estimation from the current observations. It may vary in the future depending on the distribution of other stars of Milky Way and the position of sun in it.

Sun is a huge power house that emits a large amount of power (as mentioned above) each second. It is also a huge furnace. It can absorb objects that are within its gravitational captive zone. So many things are happening in the sun. Here someone can ask, does the sun make any sound? It is really an amazing question. Isn't it? Then what kind of sounds does the sun make? All these questions have no definite answer. But it is certain that sun does make some kind of sound. Some of them are in the audible range and the rest are beyond the human audible range. Sun makes sound in various frequency ranges in the convection zone. It travels in different directions and the major part of it is reflected from the outer boundary. The part that comes out of the solar boundary is weakened due to the absence of proper medium for its propagation. The solar waves that reach the earth and the remote part of the solar system are weak enough to be audible. Scientists measure the sound intensity in the sun and around it by Doppler Effect. The study of solar oscillations and sounds is known as helioseismology. In this branch various methods are applied to study the inside of sun. From the analysis, it has been found that sun makes vigorous sounds inside it. But as the outer medium is quite rarer the sound does not reach us (because sound needs material medium to propagate). The sun as a whole also vibrates in its convection zone due to the release of huge amount of energy. It also indicates the sound must be quite furious inside sun.

The Planets

The heavenly bodies that orbit around stars are called planets. In our solar system, we know that there are nine planets. All of them are bound to the sun gravitationally. They orbit the sun and also rotate around themselves in fixed intervals (of course it is not fixed for ever rather change gradually). Looking at the whole solar system it seems that someone has programmed nicely planned game of rotation. But what is the real fact? Let us have a close look at the planetary system of sun.

No one knows, whether the planets and the sun were created simultaneously or the planets followed the sun in the time scale. Some scientists say (as we have seen above), the sun and its planets were formed simultaneously form the same nebula. Gradually, when the sun colleted more and more particles to be massive the planets could not do so due to the lack of mass in their core. Rather they collected small masses like the asteroids, cometary particles, planetesimals, rocks and dust particles. Even some small masses were captured by them as satellites. When the sun became powerful and massive enough, the smaller bodies were attracted towards the sun. Many of them got absorbed by the sun directly and some others survived due to some reasons like the collisions with other planetary bodies. After a long time they were stabilised in planetary orbits as we see today. Some others say that sun has gathered the planets, while orbiting around the galactic centre. There is also another proposal that sun had a less luminous binary and it collided to fragments and the planets were born (a kind of catastrophic theory). As it has been mentioned above, the disintegration of parts of sun become the planets, according to some scientists. There are both pros and cons of all these proposals and a better explanation will come, when we can study our sun, our galaxy and the universe more accurately and elaborately.

Mercury

After the sun its nearest member of the solar system is Mercury. It is not the hottest planet though it inherits the largest amount of sun shine from the sun. It is eighth largest (and thus second smallest) among the planets (with diameter approximately 4800 km) but a bit smaller than the moons of Jupiter (Ganymede and Callisto) and Saturn (Titan) in radius. In contrast, it is almost twice heavier than those moons. Its average distance from the sun is 57.9 million km. But Mercury's orbit is quite eccentric (presently its eccentricity is 0.206) and the eccentricity is not a constant rather it changes with the position of Mercury around the sun. In terms of distance at perihelion it is 46 million km and at aphelion 70 million km from the sun. It has also got the strange property of "3/2 resonant state of rotation" that means it changes its orientation three times in each two rotations around the sun (i.e., it rotates around its own axis in every 59 days, while it orbits the sun in every 88 days). It was one of the highly unpredictable phenomena of Mercury and, was revealed in June 2004 by Correia and Laskar. It proved that the chaotic behaviour of the solar system is not only in the outer solar system rather it is also present in the first member of the solar system itself. The eccentricity of Mercury's orbit is also quite chaotic. It varies from 0 to 0.45. Mercury's surface is highly cratered like that of our moon's. It has an atmosphere, which is very thin in comparison to ours, where helium is 95% and rest is hydrogen. There are many reasons behind the thin atmosphere of Mercury. First of all the solar winds affect its atmosphere a lot. Due to high temperature during the day time the atoms and molecules easily escape the atmosphere of Mercury (due to high temperature the molecular speed increases the escape speed of Mercury). This thin atmosphere makes the planet's sky quite dark. As the day of mercury is too long, one side is too hot (reaches around 500^0 C), while the other side freezes (as cold as -200^0 C). Mercury's core is very heavy and contains heavy metals like our earth.

Venus

After Mercury, Venus is the second member of the solar system in terms of separation from the sun and our nearest planet. It is the hottest planet, due to its atmospheric compositions and special whether conditions. It is sixth largest among the planets. After sun and moon it is the next brightest object seen to the naked eye. That's why it is known to the mankind since a long past. In many ways Venus is quite similar to our earth. Its size is quite comparable to ours and its mass is also 80% of our Earth's. Like Earth, Venus's orbit is quite circular with eccentricity 0.007. Venus takes almost 0.62 years to revolve once around the sun. But for one rotation it takes a long time almost 243 days, which is a bit longer than the Venus year. At equator its rotational speed is 6.52 km/h. Venus is quite different from the rest 8 planets from its rotational properties. It rotates in clockwise direction, while the other planets do in the opposite. Thus the tilt of the axis is $177.36^{0,}$ which means that there is no big seasonal change like in our Earth or Mars. It has no satellites like Mercury. The gravity of Venus is a little bit less than our earth (8.87). Venus has a very dense atmosphere and it is very stable like our earth's. The reason behind the stability is the slow rotational speed and considerably suitable distance from sun. The pressure on the Venus surface is almost 80 times than our average. It is due to the presence of a dense atmosphere of heavy molecules. The Venus atmosphere is very rich in CO_2 (96.4%) and the rest is Nitrogen. In the upper atmosphere some amount of sulphuric acid is there in the form of dense clouds. This richness of CO_2 makes the conditions quite terrible. A green house effect goes on in the atmosphere and it absorbs almost all the heat that it gets from the sun. Thus the average temperature of Venus is 643^{0} C. After earth Venus is the next dense planet of solar system. Venus had a large number of volcanoes on its surface and there may be some alive today. This makes the surface full of craters. Due to these huge similarities, it is thought that Venus is perhaps our future model (some day Earth will be like Venus when the temperature will rise to a high value). May be after some million years our Earth will be like the Venus. Venus transit is an interesting phenomenon, which happens when planet Venus moves before earth in between sun and earth in a straight line.

Earth

Earth is the third planet from the sun and quite different from the other planets in many aspects. Its distance form the sun is 1 AU or in terms of miles it is approximately 92 955 887.6. It is the only planet which has life on it and quite stable in terms of atmospheric and surface conditions. Earths orbit around the sun is almost circular with eccentricity 0.0167 (current value). But in fact it is not a constant rather oscillates a little about a mean value. Earth takes 23 hours, 56 minutes and 4.09 seconds to rotate once about its axis, which is known as a sidereal (the time taken to complete 360^0 rotation) day (a sidereal day is a bit smaller than a solar day). Similarly, it takes 366.2422 sidereal days to revolve once around the sun, known as a sidereal year. But it is equal to 365.2422 solar days. Earth is the third rock planet of the solar system with highest density, in the whole system. It is also the only planet known to harbour life (of course we do not know about the moons of other planets which have some chance of having elementary life if they have liquid water). To the outside space it looks blue due to the presence of liquid water. Of course the speciality of earth can be understood easily from its composition. Earth has mainly iron, oxygen and silicon as its building material. The other major compositions are magnesium, nickel sulphur, metal oxides etc. Iron is mainly present in the core. It is perhaps the main reason behind the earth's magnetic field. Oxygen has the highest contribution for the biosphere on earth. The atmosphere of earth is so balanced that the two prime elements oxygen and nitrogen make the whole thing quite perfect for the earth inhabitants. In addition to all these things, earth's surface is also quite unique. Almost three forth is covered by water and the rest is land. Below the surface and above the core is the mantle. It is thought to be the heaviest part of planet earth. There are also different sub-layers of the mantle. In terms of mass earth is the fifth largest planet of the solar system, with a mass of 5.9742×10^{24} kg. After mantle, the second largest share of this whole mass goes to the core. The mass of the water world (including oceans and fresh water) is only around 0.0234427328 percent of the whole mass. Out of this only 1 percent (actually a bit less than that) is fresh water and the sources of fresh water are being constantly fed through the water cycle processes. Similarly, the mass of the whole atmosphere is only $8.53985265 \times 10^{-05}$ percent of the earth's mass. Whatever we shall have here about earth, it will not be enough at all.

Moon

Earth has got only one natural satellite. It is our moon. It is about 3.85×10^5 km from our earth. In the past we had many mysteries about the moon. But after the historic lunar mission it is now a well known place for us. It is only second stellar body the human being has visited. Though moon is quite near to us, its origin is still not known perfectly. There are many theories about the origin of moon, but what is the actual fact, no one knows. The potential theories will be presented here and the choice is yours. Some scientists say that moon was captured by earth before a few billion years. Initially, the moon was in a heliocentric orbit around the earth, perhaps farther than earth from the sun (in an outer orbit than our earth). Due to its smaller mass and proximity to earth it gradually came closer to earth (being captured by earth's gravity) and its life was changed forever when it started revolving around our earth. At that time both moon and earth had different speeds of rotation (faster than what we observe today). Due to the interaction with both sun and earth moon's rotational speed decreased gradually and at last came to a stand still as we see today. Some others say that moon was the result of some collision with earth a long ago by a foreign body. The impact was so intense that some part of earth went out of earth and started orbiting the earth. Gradually, they were combined to form the moon. Some others say that the foreign body that collided with our earth was also captured by earth and it is what we call moon today. There are both positive and negative points of all the theories.

Mars

Now we are going to look at the red planet. It is the forth planet from the sun and seventh largest (larger than Mercury and Pluto) among the planets. As it is visible to the naked eye it is known to the mankind since a long past. Mars is quite smaller than earth and differs in many ways from earth.

Its mass is 6.6485×10^{23} kg which is just little above the 10% of earth's mass. Its eccentricity around the sun is bit more than earth's (0.093 of Mars in comparison to 0.0167 of earth). Its minimum and maximum distances from earth are respectively 55.7×10^5 km and 401.7×10^5 km respectively. But in contrast, it is also similar to earth in many respects. Obliqueness of its polar axis to its orbit is 25.29^0, which is quite nearer to our earth's (23.45^0). Mars is cooler than earth and is thought to be the future destination of the mankind when earth will be too hot to live in.

Now let us see the world of the offspring of Mars. Mars has got two satellites. One is Phobos and the other is Demios. Both are quite small in comparison to the satellites of other planets. Demios is the smallest satellite of the solar system. Phobos is a bit larger than Demios but still quite small. Both these satellites are quite nearer to the Martian surface. Both the satellites were discovered by Hall in 1877. It is predicted that these satellites are actually captured asteroids. Some other astronomers believe that they are from outer solar system. Anyway looking at their behaviour it has been calculated that Phobos will collide with mars within 50 million years. Both the satellites are highly cratered and made from rocks rich in carbon and ice like the C-type asteroids.

The Asteroid Belt

In between the orbits of Mars and Jupiter there are large numbers of small heavenly bodies, which orbit the sun in quite chaotic manner (of course not all but most of them). These small rocky solid bodies are called asteroids. In size the asteroids are quite smaller than the planets but some of the larger asteroids can be as large as a 1000 km in diameter (Ceres the largest known asteroid is about 1000 km in diameter). But most of the asteroids are only a kilometre or less in diameter. Sixteen asteroids (so far discovered) have diameters more than 240 km or more. Most of the asteroids are confined between the orbits of Mars and Jupiter. But some of them are found well within earth's orbit and stretch beyond Uranus. As the asteroids are small in size their orbits are not permanent; rather they are affected by the planets and change frequently. Thus the chaotic behaviour is common in them. Due to frequent collision among themselves they pound each other into tiny shapes and sometimes even into dust. There are many evidences of collisions with earth in the past. Asteroids can be the cause of the extinction of Dinosaurs and the Great Dying. The small asteroids very often brighten our sky as they enter into the atmosphere and start burning themselves as meteors. There are many mysteries about the origin of the asteroids. Some scientists say they were created by the collision of a planet or big satellite in between Mars and Jupiter. Actually, the asteroids are the objects that never got together as a single body. The combined mass of the asteroids together can be a heavenly body similar to our moon with 1500km in diameter (which is smaller than the radius of our moon). Only Ceres contains almost one third of the whole mass of the asteroids.

Asteroids are one of the most interesting objects of astronomical studies due to many fundamental reasons. First of all, they are thought to be created at the same time with the solar system. So, by examining the materials and compositions of the asteroids, we can get more information about the origin of the solar system and its past. Secondly, they are one of the potential threats to the earth. An asteroid of diameter around 10km is enough to demolish the biosphere from the earth. As the speed of the asteroids is quite high (normally it can be around 10km/sec when they enter the earth atmosphere) it could impact a huge momentum and energy, which is too dangerous for the earth. Thirdly, many astronomers think that there are some links between the comets and the asteroids, perhaps they have the same kind of origin and some of the asteroids are converted into comets in later life. The fourth reason is their chaos behaviour. From the asteroids the chaos nature of the heavenly bodies can be studied. The reason why the asteroids leave the asteroid belt is also one of their chaos natures. Actually when some of the asteroids are quite far from both Jupiter and Mars, they get attracted by the gravitation of sun or other planets (when their rotational inertia is not enough to maintain the old orbit) and leave their home, the asteroid belt. Due to this same reason, Jupiter absorbs a large number of such asteroids every year and protects other planets from collisions. This is similar to the cleaning of the space around Jupiter. So many astronomers call Jupiter as the vacuum cleaner of the solar system. Jupiter itself is also one of the main reasons behind the elliptical nature of the asteroid orbits. Due to its gravity the near circular orbits of the smaller bodies like the comets and asteroids get more elliptical and change their trajectories.

There are various types of asteroids according to their chemical compositions. The classification is done according to their spectral observations. The main types are C-type, S-type, M-type, E-type and R-type. Almost 75% of the asteroids belong to the first category of C-type. These are quite similar in composition to sun minus hydrogen, helium and other gaseous contents. They are also not hot rather far cold than the sun. The spectra of these asteroids are generally of blue colour and flat in nature. "C" stands for carbonaceous. The S-type asteroids are the next common type. Asteroids of this type could count up to 17% of the whole asteroid population. The letter "S" stands for silicaceous because silica is one of the main ingredients of these asteroids. The spectra of these asteroids are reddish in colour. The M-types contains mainly metals like iron and nickel. The letter "M" stands for metallic as these asteroids are metal rich.

Jupiter

Jupiter is the big brother of the planets of the solar system. It is the largest planet and the second largest body of the solar system after sun. Jupiter is more than twice as heavier than the rest of the planets of the solar system (mass of Jupiter= 1.9×10^{27} kgs). Jupiter is unlucky not to be a star due to the lack of some extra mass (actually had Jupiter been 80 times more massive it could have started the nuclear fusion by itself). It is the first member of the outer solar system and the fifth from sun. It is also the planet with too many mysterious things. It does not have solid surface rather the gaseous layers just get dense towards the centre. That is why; Jupiter is also called as the gas giant. In diameter it is 11 times larger than our earth and almost 20% larger than Saturn. The radius and diameter of Jupiter is actually calculated from 1 atmospheric pressure (where the pressure of the atmosphere of Jupiter is equal to the pressure of earth's atmosphere at the sea level) as we do not know where the solid state starts. Jupiter's orbit around the sun is almost circular with eccentricity 0.0489. It takes almost

11.9 years to orbit around the sun and its rotation around its own axis is 9.9259 hours. But do not forget that like other gas planets Jupiter too, has the differential rotation. Its inner layers and the core do not rotate at the same angular speed as we see its surface does. The temperature of Jupiter at its surface is not known exactly. It is predicted to be somewhere around -110^{0}C at the one atmosphere level. At the higher atmosphere it is less. But as we go towards the core the temperature increases. It has been found that Jupiter radiates enough energy than what it gets from the sun. So how these things are possible? This mysterious feature is explained by Kelvin-Helmholtz mechanism according to which a planet or star could produce heat inside it, when the outer pressure reduces. When the pressure is reduced and the star or planet is cooled down and contracts a bit (this is also possible due to gravitational compression, which is thought to be the main reason in the heat production in the Jovian core). This contraction can produce a large amount of heat inside the planet or star around its core. This mechanism is said to be quite active in Jupiter and thus the temperature of the core of Jupiter could be around 30 000^{0}C (which is almost 6 times higher than the solar surface temperature). Jupiter contains mainly Hydrogen gas in its atmosphere and bulk. Hydrogen constitutes almost 89.8% of the Jovian atmosphere. Helium and some other rare gases constitute the rest 10.2%. But in the buck there are many types of compounds like ammonia, methane, ethane, sulphides etc. Some amount of water vapour is also found in the Jovian atmosphere. Some water is thought to be in the bulk in the form of ice. The water in Jupiter can be due to the comets and foreign bodies, which are embedding every year into Jupiter. Some scientists say it could be from the early nebula, from which the whole solar system was created.

There is a huge red spot on Jupiter known as "The Great Red Spot" which can sallow several earths within it. It was discovered in 1665 by Cassini. It is perhaps the results of vigorous Hurricanes in that spot, where the pressure is more than the surrounding areas and thus called a super zone. The temperature in and around the Great Red Spot varies between 111 K and 125 K. The actual reason behind the Great Red Spot is not known yet. But it is predicted that the violent storms in the Jovian atmosphere could be the main reason (because very strong and powerful wave patterns have been observed around the Great Red Spot). The Great Red Spot is not a constant configuration on Jupiter's surface rather it changes with time.

Jupiter too has rings like other gas planets. The rings of Jupiter are similar to that of Saturn's but they are not that thick like that of Saturn's. So they were discovered quite late by Voyager 1 mission. There are three parts of the ring. They are named as Halo, Main and Gossamer (from the inner to outer that means halo is the innermost ring and the Gossamer is the outermost). The outer Gossamer extends almost more than three times the radius of Jupiter from its centre. The rings mainly contain dust and gaseous particles, but there could be small boulders and rocks in those rings (of course not large like that of Saturn's). Their density is quite small in comparison to Jupiter's density (mean density of the rings is around 5×10^{-6} g/cm^3). The Auroras of Jupiter are some of the Jupiter's spectacular features. The Auroras are observed clearly by the HST (the Hubble space telescope). The Auroras of Jupiter are of course fainter than that of the sun's, but quite significant. When the magnetic lines of force (or can be called magnetic flux) comes out of the North Pole it interacts with the hydrogen of the atmosphere and auroras are formed. Similar things happen near the planet's South Pole, where they enter into Jupiter. This forms the magnetosphere of Jupiter.

So far 61 satellites of Jupiter have been discovered. Thanks to the space missions and the Hubble Space Telescope. There could be even more (still undiscovered). But out of them only 4 are quite big and are well known as the Galilean satellites. They are Io, Europa, Ganymede and Callisto. They are also known as J(I), J(II), J(III), and J(IV) respectively. Ganymede is the largest moon of Jupiter. Io has many features which show that it could be one of the suitable destinations to harbour life. Many of the smaller satellites are actually the asteroids captured by the gravitation of Jupiter and they can not live very long like other stable satellites.

Now at last we shall have a look at the cleaning operations of Jupiter that we have mentioned in the previous section of asteroids. Jupiter is the second heaviest object of the solar system and also the second highest object absorber. It absorbs a large number of asteroids and comets every year. If the object is small we cannot see it. But when a large comet or asteroid is absorbed its collision with Jupiter, before the complete captive is very attractive. The collision with Shoemaker-Levy 9 comet is one of such examples. In 1994 it collided with the gas giant several times in a spectacular event of the century.

Saturn

Saturn is the second largest planet in the solar system and sixth from the sun. It is also quite different from its appearance from its outside due to the presence of thick colourful rings. It is also a gas planet and has got a lot of satellites like Jupiter. Its mass is 5.68×10^{26} kg and the average distance from the sun is 9.54 AU. The orbit of Saturn around the sun is not very elliptical but in comparison to our earth's orbit is more elongated with eccentricity 0.0565. Saturn takes 29.46 years to orbit the sun. But its rotation about its axis is quite fast (one Saturn day = 10 hours 39 minutes). Due to this fast rotation (also due to the presence of gases and liquids in large amounts) Saturn is quite flattened at its equator (equatorial diameter is 120536 km and the polar diameter is 108728 km). As Saturn is quite far from the sun its average surface temperature is 150K (-139^{0}C). But at the bulk it is very hot (around 12000K). Like Jupiter Saturn too does not have a distinction between surface and atmosphere. Its atmosphere mainly consists of hydrogen and helium, but in the inner part there is trace amount of water, methane and ammonia in liquid state. At the centre perhaps there is a rocky core. As the pressure increases towards the centre from surface the gases are found in their liquid forms. The central temperature is due to the Kelvin-Helmholtz mechanism. According to this mechanism the gas planets are contracting due to their own gravity and this compression produces a significant amount of heat. Now let us see at those beautiful rings around Saturn. The rings of Saturn were for the first time seen by Galileo in 1610. But in 1659 Christian Huygens calculated the details of various rings. Though it is a common feature of gas planets to have rings around them the actual reason and nature of the rings are not known yet. There are many layers in Saturn's ring and their colour, thickness and brightness etc are also different. The main potential reason behind Saturn's ring is the collision of a gas satellite. Perhaps before several million years ago a gas satellite of Saturn got close to it and finally merged with it. But its gases started forming ring in this gradual process. The compositions of Saturn's rings are also different. But the major parts are water ice particles of different dimensions. The whole ring system has been divided in to five sections. They are G, F, A, B, C in the order from out side to inside. In fact each of these ring sections has thousands of sub rings or ringlets. Out of these, A, B and C are clearly visible while G and F are quite faint. There is a large gap between the A and B sections. It is known as the Cassini Division. From Voyager mission it has been found that the Cassini division also contains a large number of faint rings. The thickness of the rings can be around 10km and the whole mass is around the mass of our moon. These rings revolve around Saturn at different speeds known as differential rotation. In addition to the ring structures there are also spoke-like structures in the rings in the radial directions. Spokes cannot be possible in the gravitational fields where differential rotation exists. So there may be some other forces like electromagnetic forces, which keep the particles in a straight line in spokes.

Now let us have a look at the moons of Saturn. Like Jupiter, Saturn too has a large number of satellites. Due to the advances like HST and many other powerful telescopes we are now able to see a lot of smaller bodies around the planets. The first initiative of discovering the satellites of Saturn was successful with the Voyager missions. The actual number of planets of Saturn is not known. It could be more than 50. But as per the 2004 findings there are 33 satellites of Saturn (according to NASA). Their size ranges from very big to very small. Out of all those satellites only 7 satellites are big enough and known as the Major Satellites of Saturn. They are Titan (also known as SVI to the scientific community and having a mass of $1.345.5 \times 10^{23}$ kg, radius 2,575 km and mean density 1881 kg/m^3), Rhea (also known as SV, with mass of 2.31×10^{21} kg, radius 764km and mean density 1240 kg/m^3), Iapetus (also known as SVIII, with a mass of 1.59×10^{21} kg, radius 718 km and mean density 1020 kg/m^3), and their smaller cousins Mimas (SI), Enceladus (SII), Tethys (SIII), Dione (SIV) and Hyperion (SVII). Titan is the most important satellite of Saturn for the scientific community, as it is quite large and may be suitable for future human habitation.

Uranus

Uranus is the seventh planet from the sun and is the third largest in diameter (equatorial radius 25,559 km). It is also the fourth largest planet after Jupiter, Saturn and Neptune. Its mass is 8.68×10^{25} kg and almost 14.5 times heavier than earth. Its mean distance from the sun is 19.218 AU. The solar orbit of Uranus is not that elliptic with eccentricity 0.0457. Uranus is different from other planets in the sense its central axis is almost parallel to its orbit (where as the other planets have near perpendicular axes). Uranus is quite far from sun, so its temperature is as low as 58.2K or -214.95 ^{0}C. Looking from earth through a common telescope it does not seem that Uranus has any ring system. But closer investigations by space crafts and powerful telescopes reveal that Uranus has a ring system like other gas planets. Of course they are not that large like that of Saturn's. The equatorial radius of Uranus is 25,559 km (measured at 1 bar level, which is quite larger than the polar radius), where as the radius of its outermost ring is 51,149 km. It shows that the outermost ring is more than double the equatorial radius. But the reason why it is not visible to small telescopes is, it is not that dense. Interesting thing about the rings is that they do not revolve at the same speed as the equator and some of the outer rings do it in elliptical orbits. The rings are made up of ice and small rocks. The rings are not that colourful and look a bit dark. The brightest ring is known as the Epsilon ring and it is the outermost. Uranus is blue in colour due to the presence of methane in its atmosphere. Methane gas filters out the red part from the solar spectra and reflects the blue colour. Uranus takes almost 84 years to complete one revolution around the sun. But its rotation about its own axis is too fast and one Uranian day is only 17 hours long. Uranus is mainly made of various gases and their ices. The bulk of the planet contains around 15% hydrogen and a trace amount of helium and methane. Other constituents of the planet are not known. Its core does not have rocky nucleus like that of Saturn and Jupiter. Its atmosphere contains 85% hydrogen, 13% helium and 2% methane.

Uranus has 26 satellites, out of them 21 have been named and the rest 5 are yet to be named. Most of the satellites of Uranus were discovered by Voyager 2 space craft. Now with the help of some ground based telescopes and the HST some more satellites have been discovered. The main satellites are Titania (also known as UIII), Oberon (UIV), Ariel (UI), Umbriel (UII) and Miranda (UV). Titania is the largest satellite with mass 3.52×10^{21} kg and radius 790 km.

Neptune

Neptune is the eighth planet from the sun and the third largest (by mass and fourth largest by size) in the solar system. Of course it is smaller then Uranus in size (radius 24,764 km and mass 1.0243×10^{26} kg). It is also in a far cold zone. It is also known as blue giant as it looks blue and a huge gas planet like the three other outer solar system gas giants. Neptune is thought to be composed of ice and rocks. Its atmosphere contains mainly hydrogen and Helium with a little amount of methane. It is predicted that the blue colure is due to the presence of methane which absorbs red colure and reflects blue colour. But the blue colour is too intense. So it may be due to some unknown phenomenon. Neptune too has a ring, but it is not very dark like that of Saturn's. Neptune has a big dark spot like that of Jupiter's Great Red Spot known as Great Blue Spot. It was actually found by Voyager-2. But recent observations by HST suggest that it is no more there. Neptune's winds are perhaps the strongest in the solar system. Sometimes they reach speeds as high as 2000 km/h. Due to its large distance from the sun, Neptune is very cold. Its temperature is generally less than -200^0 C. The coldest temperature in the solar system is -230^0 C, recorded on the moon of Neptune, Triton. Observations suggest that the moon Triton is getting closer to Neptune and someday, it may merge with Neptune forming a nice spectacular ring around Neptune. It is funny that Neptune is now the most distant planet from the sun and will remain so till 2008. It happens due to the highly eccentric orbit of Pluto.

There are 13 known satellites of Neptune. Triton is the largest among them with mass 2.14×10^{22} kg and radius 1,353.4 km and mean density 2,050 kg/m^3. There are 7 middle-sized satellites and the rest five are quite small (thought to be captured from the Kuiper Belt). Triton is the most studied satellite of Neptune. It has some characteristics of harbouring life, and probably, will be a suitable place for the human being in the future.

Pluto

Pluto is the farthest member from sun (of course not now till 2008) and also the smallest among all the planet members of the solar system. It is also smaller than the satellites of some planets (around half the size of moon). Its average distance from the sun is 39.5 AU. Its orbit is highly elliptical with eccentricity 0.249. Its direction of rotation is different from the rest 8 planets (tilted at an angel 17.14^0 to the common plane) and thus it is perhaps from outer solar system. Its plane of rotation is also different and inclined to the common plane of rotation of the other 8 planets. It takes 248 years to orbit once around sun, out of it for 228 years it is the farthest from the sun(for 20 years it is nearer than Neptune). The orbits of Pluto and Neptune intersect each other and also very close to each other. Thus some astronomers say that perhaps Pluto was a satellite of Neptune previously. But now the two planets do not come very close to each other as their orbits are in resonance (for each 2 orbits of Pluto, Neptune does 3 orbits around the sun). So there is no chance of collision between Neptune and Pluto in the near future. Pluto is the only planet to which there have been no space missions. So it is the most undiscovered planet in the solar system. Whatever data is there about Pluto are just smart predictions or based on the facts of HST (Hubble Space Telescope). Surface compositions of Pluto are H_2O, N_2, CH_4 and solid CO etc. Its atmosphere contains mainly N_2, CH_4 and CO. The surface temperature of Pluto is around 40K or -233^{0}C. Due to this low temperature Pluto's atmosphere exists mainly when it is near the sun or in its perihelion. Pluto's surface gravity is too small in comparison to ours (1/15 of earth's gravity). So the gas from Pluto's atmosphere escapes when they are in the gaseous form.

Pluto has a satellite which is quite large in comparison to the satellite and its primary (i.e., planet). Its name is Charon and it was discovered in 1978 while observing Pluto. The transits of Pluto and Charon are really nice to see. Charon takes 6.39 days to orbit once around Pluto, which is also equal to the spin periods of both Charon and Pluto. Charon and Pluto are quite similar to Neptune's moon Triton and these three are quite different from the rest of the solar system. So it may be true that these three bodies may be from outside solar system or they are from the remote solar system and, were divided into three pieces as the result of some collision with some planet/planets.

In this chapter the overall summaries and most significant facts of the planets of the solar system have been given. As it has been tried to represent the whole things in a concise manner it is not possible to look at every detail of the planets. Going in to the deep you can learn the whole story of the solar system (some of the rich sources have been mentioned in the references at the end). In the coming sections of this chapter we shall look at the post Pluto regions of the solar system.

After Pluto starts the chaotic zone. Of course chaotic behaviour is a part of the whole solar system. Even in sun and Mercury there are some chaotic behaviours. Between Mars and Jupiter's orbit the asteroid zone is also full of particles having randomness. So the in the whole solar system there are a lot of disorders within order. But after Pluto it is highly disordered space. Very few objects in that zone have ordered motion. This zone is called Kuiper belt because it has been found that they have formed a ring around the sun (which was first proposed by astronomer Gerard Kuiper, who discovered that Titan has an atmosphere and in the past it was similar to ours). Actually, the Kuiper Belt starts after Neptune. So sometimes the Kuiper belt objects are called as the trans-Neptunian objects. The 9th planet Pluto is also a Kuiper belt object. There have been found a lot of small asteroid and comet like bodies, which orbit the sun in this region. There are a lot of curiosities about this zone due to two main reasons. First of all it is the zone where the comets are thought to be originated and thus it will be quite informative about the comets and other chaotic objects in that zone. Secondly they are thought to have formed at the same time as the solar system. So it can give a lot of information about the origin of our solar system, other stars and galaxies.

Kuiper Belt

The first Kuiper belt object was discovered in 1992 (after 19 years of the death of Kuiper) by David Jewitt and Jane Lulu. Soon after that they found more objects in that region. Now very large objects have been found in the Kuiper belt. Some of them are even bigger than Pluto's satellite Charon. Many researchers are saying them as planet X. But actually due to their small size they have not been included in the planet list. Even some astronomers do not accept Pluto as a planet. There are more than 100 objects having diameter more than 100km have been discovered. Many of them form 2:3 resonances with Neptune like Pluto. So they are thought to be the smaller versions of Pluto and known as Plutinos. The largest among them are Sedna, 2004 DW, Quaoar, Ixion, 2002 AW197 and Varuna etc. Out of these Sedna (also known as 2003 VB12, its radius is around 900 km and mass little less than that of Pluto) and 2004 DW (its radius is around 1600 km and mass is larger than Charon) are bigger than Pluto's moon Charon. So now it is a question that whether our Pluto should be given the status of a planet any more? There are many interesting observations in the Kuiper belt. There are some binary objects in the Kuiper belt, which orbit each other.

The actual boundary of the Kuiper Belt has not been known yet. But it is predicted that the main active Kuiper belt is up to 100 AU or around that from Neptune's orbit. But there is a chance of scattered objects at a distance of 1000AU where the objects orbit the sun nearly in a circular orbit. That means the Kuiper belt starts after Neptune's orbit and finishes with the boundary of the solar system. Whole study of the dynamics of the Kuiper Objects will give a good idea about the region.

The Boundary of the Solar System

We have gone through the major facts of the solar system. Now let us conclude the study of the solar system with the solar system's boundary. Boundary of the solar system is not known perfectly. But it is assumed that, where the solar wind meets the interstellar gas and dust particles, is the boundary of our solar system. The problem here is that we do not know exactly where the solar wind is touching the interstellar space. The spacecraft Voyager-1 has almost travelled more that 14 billion km (speed of Voyager-1 is 3.6 AU per year) so far. Many scientists say that it has already crossed the boundary of the solar system and some others say that it is well within our solar system. The boundary of the solar system which is known as the "termination shock", is the place, where the solar wind dissipates in the interstellar space. At termination shock the supersonic solar winds slows down to subsonic speed and energise the boundary. So the magnetic field at the boundary must have been quite strong. The space after the boundary of our solar system is known as heliopause. The other problem in detecting the boundary for the solar wind point of view is that Voyager-1 has already broken its detector (through which it could have measured the solar winds directly) in 1990. It is expected that Voyager space crafts can give us a clearer idea about the heliopause around 2020. It is because by that time they must have been in an interstellar space. Wherever the boundary of the solar system may be, the average separation between the stars around our solar system has been found to be around 1 parsec. So the influence of our sun is limited within 0.5 parsec or less than that (of course it depend on the properties of the neighbouring stars).

2

Our Nearest Neighbouring Stars and Their Stories

There are a large number of stars around our solar system. With naked eye we see only a few thousand in the night. In the day time when the sun comes to our horizon, other stars are not visible due to the brightness of sun in the local sky. We see a lot of light spots of different intensities in the night sky. Most of them are stars and galaxies, a very few of them are planets and satellites which are quite nearer to us. In the past there was no good method or technology to measure the distances of stars. So it was quite difficult to know which stars are the nearest. But, as the different methods were discovered and good observatories were built for star gazing, astronomers and physicists calculated the distances of various stellar objects. For short distances the trigonometric parallax methods (details can be found in appendix-A) are extremely helpful and accurate in measuring the distances of various stellar objects. The distances can also be measured from the telescopic observations from earth. But it may not be very accurate. Voyager space crafts were designed to calculate the parallax data and to send them back to earth. Finding the utility of Voyager space crafts, scientists designed a special satellite for the measurement of accurate distances of stars. That is Hipparcos geostationary satellite (lunched in August 1989 and its service was terminated at the end of 1993). Initially it was not that effective due to its own instability in the geostationary orbit. But after some adjustments, it was very successful in calculating the distances of various stars. The distances of the winking stars or Cepheids (they are used as references for distance measurement in the space) were previously measured by their brightness. But the Hipparcos mission found that it was wrong and the Cepheids are even farther and more luminous than the previous estimations. Now the distances of many stars within the 50 light years and beyond have been calculated accurately with the help of Hipparcos data. More Cepheids have also been discovered, which helps a lot in the measurement of stellar distances.

The Alpha Centauri System

Sun is our nearest star. After sun the neighbouring stars of the Alpha Centauri system are the next nearest stars (star system) to us. It is a very beautiful three-star system with a binary. The name Alpha Centauri came from the Past name of "Kenturus". This indicates that the star system is not the recent discovery rather it was known to our forefathers a long ago (around 7th century or even before that). Alpha Centauri system has three members whose names are Alpha Centauri A, Alpha-Centauri B and Alpha-Centauri C. Among them Alpha Centauri C is the nearest star to us. So it is also known as Proxima Centauri or Proxima. Alpha Centauri A and B form a binary system and the separation between them is about 23.7 AU (mean separation). However the distance between these two stars is not constant as they form a binary with highly elliptical orbits (eccentricity = 0.519). In terms of light-time they are around 3.45 light hours from each other. But Alpha Centauri C is quite far from Alpha Centauri A and B, and the separation between them is around 13,000 AU or 0.205567 light years. Proxima Centauri is 4.22 light years away from us. Due to this large separation astronomers are not sure that whether, Alpha Centauri C is really bound with the other two stars gravitationally or not? Though Alpha Centauri C is the nearest star it is not visible to naked eye due to its small size and low brightness (luminosity). But both A and B are quite big and bright. Alpha Centauri A is a G2 star and B is a K1 star (more about the spectral types can be found in appendix-C). Both of them are quite similar to our sun. But Proxima is quite different from all respects. Its mass is only around 12.6% of our sun's mass and the radius is 14.7% of solar radius. If it is gravitationally locked with the A and B system then its period of rotation must have been as long as half a million years or more.

Alpha Centauri A is very much similar to our sun. Its mass is a little bit more than that of sun (1.09 to 1.11 times that of sun). It is a yellow G2 V star whose surface temperature is almost same as of our sun i.e., 5800^{0}C. Radius of Alpha Centauri A is almost 1.2 times that of our sun. It is almost 150% brighter than our sun (if the brightness of sun is 1, then Alpha Centauri A's brightness is 1.543). Its age is around 5 to 6 billion years. It is 4.35 light years from our sun. The revolution period of the A and B system is almost 80 years and during one complete revolution the distance between the two stars vary between 11 to 35 AU. Alpha Centauri A is well known as Rigil Kentaurus or Rigil Kent. It is one of the brightest stars of the Centaurus constellation (the group of stars that form some kind of figure in the night sky, and have been named accordingly). As it is a big star like our sun it is expected that it should have some big planets. But so far no such planets have been observed.

Alpha Centauri B is a bit smaller and different from sun. It is the partner of Alpha Centauri A in their binary system. Alpha Centauri B is an orange star with spectral type K1 V. Its surface temperature is around 5300K. Its mass and radius are respectively 90% and 80% of our sun's. But in contrast, its brightness is only 44% of solar brightness. It is believed to be of same age as Alpha Centauri A. There is no evidence of any existing planets of this star. But some astronomers believe that there may be some planets orbiting the common centre of mass of the stars outside the binary orbit of the stars.

Alpha Centauri C or Proxima is quite different from its other two partners. It is quite dim and thus not visible to naked eye, though it is the nearest star (4.22 light years) to earth (other than sun). That is the reason it was discovered in 1915 (quite late in comparison to its partners). It is a red dwarf star with M5 Ve spectral type. Its surface temperature is as low as 2700K with brightness only 0.00006 (or 0.006%) as that of sun. Its age is not known perfectly and many scientists say it is a young star while some others say its age is same as that of Alpha Centauri A and B. Either option has some pros and cons. If it is a young star then, may be on its way to be a big star and in the future it will be brighter by consuming its fuel at a higher rate. If it is a contemporary of Alpha Centauri A and B then, it perhaps spent almost all the fuel available to it or the process of fuel consumption is too slow. The variations in the orbit of Proxima suggests that there is a companion of it with mass around 80% that of Jupiter's. But so far no such planet or brown dwarf has been discovered.

Besides being our nearest stars there are some other reasons like the high probability of existence of life make the Alpha Centauri system quite interesting to the scientific community. The first reason why intelligent life may be possible in Alpha Centauri is that it is quite similar to our sun. Its temperature is almost same as that of sun. Its brightness also quite suitable and water probably is in some of its planets, if it has some. Out of the three stars of the system, Alpha Centauri A is the most probable place for intelligent life. Alpha Centauri B has less probability and Proxima has no chance of harbouring life.

Beyond Alpha Centauri

Apart from the Alpha Centauri system, stars which are quite nearer to our earth are Barnard's Star, Wolf 359, Lalande 21185, Sirius A or Alpha CMa A, Sirius B, Luyten 726-8A, Luyten 726-8B, UV Ceti, Ross 154, Ross 248,Epsilon Eridani, Luyten 789-6, Ross 128, 61 Cugnus A, 61 Cygnus B, Epsilon Indi etc. Out of all these stars the brightest (within 20 light years) is Sirius A or Alpha CMa A. It is quite far from us in comparison to Alpha Centauri system, but due to its big size and high brightness it looks very big and bright.

In terms of distance, after the Alpha Centauri system the next nearest star to us is Barnard's star. But it is quite dim and cool star like Proxima with surface temperature around 3200K. Similar to Proxima, its spectral type is M3.8 Ve. It is almost 6 light years away from our earth. It can be seen in the northern part of the Ophiuchus constellation. It was discovered by Barnard in 1916. It is one of the fast moving stars, which changes its position by 10.3 arc seconds every year. If it continues at this rate, after some thousand years (with a good estimation it could be around 8000 years) it will be our nearest star (as it is approaching towards sun).

The next nearest star is Wolf 359. It is flare star and can be seen in the east central direction of the Leo constellation. It is sun's dimmest stellar neighbour (dimmer than Proxima and Barnard's star). It is a red dwarf with spectral type M5.8 Ve. Wolf 359 is around 7.8 light years from us. Max Wolf discovered this faint star in 1880 with the help of photographic plates. Its mass could be around 9 to 13% that of our sun's and diameter is around 16 to 20% of solar diameter. But in contrast, its temperature and luminosity are quite negligible. The variations of the path of Wolf 359 suggest a nearby companion within a radius of 1 AU. But so far there has no such partner been discovered.

Lalande 22185 system is the next nearest system to our earth. Its distance form our sun is 8.3 light years. The exact location is in the south eastern corner of the Great Bear or Ursa Major constellation. It was discovered by Jerome de Lalande in 1795. Its spectral type is M2.1 Vne which means it is a cool, dim main sequence red dwarf. Its mass is 46% of solar mass but as far as its planets are concerned it has perhaps 2 or more Jupiter like planets. The planets Lalande 21185b and Lalande 22185c are quite big and cold objects. Their approximate orbital radii are respectively 2 and 10 AU. Lalande 21185b is similar to Jupiter with mass 90% of Jupiter's and Lalande 22185c is larger than Jupiter (1.6 times). Lalande 21185 it is thought to be one of the oldest neighbours with age around 10 billion years.

Sirius A is the brightest star after sun (for us) and is the fifth nearest to us with a distance of 8.6 light years from us. It is also known as "Dog Star" and one of the oldest known to the mankind. It is also the brightest star of Canis Major and is seen in the northern hemisphere. In the past the star gazers had found some irregularities in the motion of Sirius A and in doubt that perhaps it has a binary partner. At last in 1862 it was found to be true that another faint star Sirius B forms a binary system with Sirius A. The separation between the two binaries is as close as 20 AU (of course it is not a constant rather varies with their position from 8.1 Au to 31.5 AU). Sirius B is a white dwarf with spectral type DA2-5 (D stands for dwarf), while Sirius A is a main sequence dwarf star with spectral type A1 Vm. They revolve each other in a highly eccentric orbit with eccentricity 0.59 in 50.1 years. Though Sirius B is a faint star (luminosity less than one 360th of sun) it is massive enough to affect its companion and both the stars swing due to their huge gravity.

Luyten 726-8A is around 8.73 light years away from our sun and it is the sixth nearest star (actually star system) to us. It is also known as BL Ceti as it is a variable star. This is a small, dim and cool star with spectral type M 5.6Ve. Its mass is just around 10 percent of the solar mass with a 14 percent solar diameter. So it is one of the dimmest neighbouring stars (luminosity is just 0.006 percent that of sun). This is the reason why it was discovered so late (in 1949 by Willem J. Luyten) despite being one of the nearest stars. Luyten 726-8A is not a solo star; rather it has a partner in their binary system. Its binary is Luyten 726-8B (but it is well known as UV Ceti). This is also a small, dim, cool and flare star with spectral type M6.0 V. It is almost of the same size as its partner. However, its luminosity is less than that of Luyten 726-8A (around 0.004% of solar luminosity). But in contrast, it can enhance its visibility by up to 5 times due to some transition inside it. Many times it has been observed that UV Ceti changes its brightness and orientation within few intervals. These two stars orbit in their elliptical orbit in each 26.5 years. The orbit is highly elliptical with eccentricity 0.62. But it is strange that their mean separation is just 5.5 AU. Both the stars are red dwarfs and can be seen by a telescope in the south-western part f the Cetus constellation.

Ross 154 is the next nearest star to us with a distance of 9.71 light years from us. It is too dim to be seen to the naked eye. It is a small cool star with spectral type M3.5V. It was discovered by F. E. Ross in 1925. It has almost 17 percent of the solar mass and in size its diameter is 24 percent of the solar diameter. However, it is quite faint and dim with only 0.0045 percent solar luminosity. It is a flare red dwarf which can be seen in the eastern part of the Sagittarius constellation with the help of a telescope. Ross 154 is also a variable star with reference number V1216 Sagittarii.

Ross 248 is a main sequence red dwarf star situated at a distance of 10.32 light years from the earth in the direction of the Andromeda galaxy (also in the Andromeda constellation). But it is too small and dim to be seen to the naked eye. Its spectral type is M5.5Ve. Its mass is around 25 percent of the solar mass, which is confined in a diameter which is just 7 percent of the solar diameter. However, its luminosity is something around 0.0012 percent of the solar luminosity and thus it was discovered quite late in 1925 by Ross. It is a variable star with reference number is HH Andromedae and the variations suggest the presence of a companion, which is too faint or may be a brown dwarf or a planet like Jupiter.

Epsilon Eridani is around 10.69 light years from us. It is one of the orange-red main sequence dwarf stars with spectral type K2 V. It is smaller than sun with around 85 percent of the solar mass in a diameter which could be around 80 percent of the solar diameter. While its luminosity is quite low (only around 27 % of the solar luminosity). Thus it is predicted to be a very young star (may be less than 1 billion years). But it is quite enriched in heavier elements, which indicates that it could be one of the brightest stars in the future. It can be seen in the north-eastern direction of the Eridanus constellation.

Luyten 789-6 is also referred as Luyten 789-6 A and as Luyten 789-6 A1. This star is a main sequence red dwarf with spectral type M5.5 V. It is around 10.87 light years (some other astronomers have estimated the distance to be 11.26 light years) from our earth. Its luminosity is around 0.01 percent of that of our sun. It is quite smaller than our sun. It is perhaps a part of a binary star system. Its high proper motion was first observed by W. J. Luyten. This star can be found in the central part of the Aquarius constellation.

Ross 128 is another star discovered by Frank E. Ross in 1925. The distance of Ross 128 is around 10.87 light years from our earth. Its proper motion is 1.361 arcsecs/year, which is quite significant. This was one of the main features of the star that attracted the discoverer Ross who wrote about it in a paper called "Second List of proper Motion Stars" in 1925 in the Astronomical Society magazine. It is a faint main sequence red dwarf with spectral type M4 V (some others suggest it to be bit different). Its mass has been predicted to be around 33 percent of the solar mass with a diameter which is just one tenth of the sun. But in comparison its luminosity is just around 0.0035 percent of the solar luminosity. That's why it cannot be seen to naked eye. However, it can be seen in the north-eastern direction of Virgo constellation with the help of a telescope. It is also a variable flare star.

61 Cygnus A is the next nearest star to us which is at a distance of 11.359 light years from earth. It is an orange-red main sequence dwarf star with spectral type K 3.5-5.0 Ve. In comparison to sun it is a small star with 70 percent of the solar mass and 72 percent of the solar diameter. While its luminosity is around 8.5 percent of the solar luminosity. It is predicted to have heavier elements like sun and other big stars. It is one of the variable stars, which has quite large proper motion in comparison to the other stars. That's why it was named as the "Flying Star" by Giuseppe Piazzi in 1792. It is not a sole star rather has binary, 61 Cygnus B. It is an orange-red main sequence dwarf star with spectral type is K 7.0 Ve. This is a bit smaller than its partner 61 Cygnus A with mass around 63 percent of the solar mass confined in a spherical volume having (equivalent to) 67 percent of the solar diameter. But in Contrast its luminosity is just around 4 percent of the solar luminosity. Both stars orbit in an elliptical orbit with eccentricity 0.4 (which indicates a highly elliptical orbit). Their mean separation is around 86 AU and the orbital period is around 660 years (some others say it is around 722 years). Both these stars can be seen in the south central part of the Cygnus constellation.

Epsilon Indi is around 11.82 light years from our earth. It is one of the fast moving stars with proper motion 4.7 arcsecs/year. Epsilon Indi is one of the main sequence orange-red dwarf stars with spectral type K5 Ve. It has around three fourth of the solar mass in around the same proportion of the solar volume. However its luminosity is just around 17 percent of the solar luminosity. It is one of the younger stars, whose age is in between 1 to 2 billion years. But it has perhaps more heavier metals within it in comparison to its age. It can be seen in the north-eastern boundary of the Indus constellation (also fifth brightest star in Indus). It has been found to have a brown dwarf companion.

After Epsilon Indi the next nearest star to us is the Procyon system, which is the brightest system in the Canis Minor. The Procyon system is a binary star system with a distance of 11.4 light years from our sun. The mean separation between the two stars in the binary system is around 16AU. The larger star of the system Procyon A, is a main sequence yellow white dwarf star with spectral type F5 IV-V. It is the eighth brightest star in our night sky. It is larger than our sun in both size (around 1.3 to 2 times the diameter of our sun) and mass (1.5 mites more massive than our sun). In brightness and temperature also it is quite ahead to our sun (around 7.5 times brighter). Thus it radiates more in the high energy (thus high frequency) regions. It is also a metal rich star and the heavy elements are in more amount than sun. Procyon A is thought to be a young star and is consuming its hydrogen much faster than our sun. Its companion Procyon B is a white dwarf star with luminosity much less that Procyon A. Its spectral type may be DA-F or A-F VII. But in contrast the temperature of Procyon B is higher than its brighter partner (the surface temperature of Procyon A is around 6500K, while that of Procyon B is around 8700K). It is due to Procyon B's huge mass (around 60% of our sun's) concentrated in a small volume (diameter around 2% of our sun's). It is thought that the Procyon B was a bright star before many years and after burning out all its fuel, it has been turned in to a white dwarf and cooling down since a long time. From these observations it is sure that Procyon B is much older than Procyon A.

The next nearest star system are the binary of Sigma 2398 system, which are well known as the Struve's star because the separation between the stars was first calculated by Struve. The average separation between the binaries is around 56AU (with semi major axis 42 AU). Struve 2398 A is perhaps a main sequence red dwarf star with spectral type M3.0 V (in some catalogues it is also registered as K5). It is the larger one in the binary system and has only 36% of the solar mass in the volume of 54% of solar diameter. The two stars orbit in an elliptical orbit but the characteristics of the ellipse has been found to be variable. The other star of this system Struve 2398 B is a flare star with spectral type M3.5 V. It has 30% of solar mass with the 55% of the solar diameter. These two stars are not visible to naked eye but can be visible by a small telescope in the Draco Constellation.

Groombridge 34 binary system is the next nearest star system to our sun. It is 11.6 light years from our sun. It is in the Andromeda constellation and can be found in the northwest of Andromeda galaxy (M31). Both the stars of this binary system are quite small in size and mass. The larger companion Groombridge 34 A is a main sequence red dwarf star with spectral type M1.5 Vne. It mass and radius are only 38 and 34 percent of our sun's. The separation between the stars is around 147 AU and their orbit is quite circular with eccentricity quite nearer to 0. But some other calculations show the orbits are quite elliptical. The other partner of the system Groombridge 34 B is also a main sequence red dwarf star with spectral type M3.5 Vne. It has only 8% of the solar mass in the 19% of the solar diameter. This system is dim and faint to be seen by naked eye.

Lacaille 9352 is 10.7 light years from our sun. It is a cool and dim star. Its spectral type could be M 1.5 Ve. As it is a quite faint star its properties are not known exactly. Its mass is perhaps 47 percent of our solar mass and the diameter could be in the 46-58 percent range. In comparison to this its luminosity is quite less just around 1.1% of our solar luminosity. It is found in Constellation Piscis Australis.

Tau Ceti is located 11.9 light years away from sun. It is visible to the naked in the night sky in the south central section of Constellation Cetus. Of course it is cooler and smaller than our sun. It is a yellow-orange dwarf with spectral type G8 V. It has around 80% of the solar mass in the 77% of the solar diameter. But it is an old star and running out of hydrogen and quite rich in heavier elements. It is the nearest sun like star, which has a large number of asteroids and comets. But the number of these chaotic objects is almost 10 times higher than our solar system and they frequently collide with each other and Tau Ceti as well. So this discards any possibility of life in Tau Ceti. Dust layers also have been discovered around Tau Ceti, which are quite similar to the dust layers found in the solar system.

BD +5deg 1668 is a red dwarf star with spectral type M4. It is at a distance of about 12.3 light years from us. It is faint dwarf star and thus cannot be seen with naked eye. It is also known as Luyten's star. The name BD +5deg 1668 seems to be too odd. But the scientists and astronomers who discover them, name them according to their own name or the location of the stars in the sky (sometimes according to the constellation).

Luyten 725-32 is around 12.4 light years away from the sun. Its spectral type is perhaps M5.5 V. It is a faint, dim and cool star. It is quite smaller than sun in both mass and size. Its luminosity is just around 0.0188 percent of the solar luminosity. So it is not visible to the naked eye.

Kapteyn's Star is situated around 12.7 light years from sun in the Constellation Pictor. It was discovered by Kapteyn in 1897. It is a dim red sub dwarf with spectral type M1 VI. It is much cooler and fainter than sun. Thus it is not visible to the naked eye. After Bernard's star it is the second fast moving star, whose motion has been recorded to be changing 8.7 arcsecs per year.

Kruger 60 A and B are the two stars in a binary system with almost 13.05 and 13.06 light years respectively from us. The spectral type of Kruger A is M3V and quite faint and thus seems quite dormant. Its luminosity is 0.01 times that of sun. Its parallax is 0.250 arcsecs. Kruger B is even more faint and its luminosity is quite less only 3.4×10^{-4} times that of sun. Its parallax is almost same as that of Kruger A.

The data and figures given for the above nearest stars may change in the future as the advanced technology and science yield better and more accurate results. Similarly, new Cepheid variables are being discovered in different parts of our galaxy. The new calculations according to their locations give some corrections to the old calculations. Stars also move with respect to each other. So the star, which is nearest today, may not be the nearest star after 1000 years. As we have seen in case of Bernard's star which may be our nearest star after several years. More than that the discovery process is going on and new stars are discovered and some of them even nearer than the well-known stars. Here we have given some data about our nearest stars. But as we go deep into the space the number of stars increase. The distribution of stars, in our galaxy is also not uniform. Rather in star clusters and inner arms of the galaxy there are a lot of stars, but in between the spiral arms there are very few stars. Similarly, there are areas where new stars are being born and at some different places old and aged stars are exploding as supernovae.

Various Stages of Stars

Here while looking at the details of our nearest neighbouring stars, we should look at the life span, various stages and the other fundamental facts of the stars. The stars are born from different cosmic processes. Most commonly they are born from the vast mass of gases and dust particles. Here you should be clear that the gas means generally a mixture of hydrogen and helium, dominated by hydrogen, and dust means, the mixture of smoke sized particles. It can be any mixture but most probably composed of ice carbon and silicates. Due to the effect of gravity they start accumulating around heavier centres. As the mass increases gradually, the process of sucking the dusts and gases also increases. In this process the gases and dusts come towards the centre at a high speed which in return increases the temperature due to friction. After crossing a critical limit of mass, the temperature rises to a very high value at which something different starts to happen. That is the starting of nuclear fusion of hydrogen nuclei into helium nuclei. In this process, the temperature, brightness and energy increases incredibly. That bright body can be called a star.

The stars live and end their lives according to their fates and performances. That means their life span depends on their mass and matter they have. It also depends on the rate at which they consume their fuel and their surroundings. If a star is burning faster, then its age becomes less, but it becomes very bright and luminous. Below some details of different forms of stars and their life processes have been given.

Main Sequence Stars

Main sequence stars are those, which are consuming the hydrogen for their existence. Their brightness is due to the lower order fusion reaction. In this process they continuously convert hydrogen into helium and, can pursue that for a long time. The amount of Hydrogen, in main sequence stars is found in huge quantities (between 30 to 90 % of the whole mass or even more than that depending on the age). The main sequence stars are generally young stars. As they age, they burn up their fuel and may go for a post-main sequence phase (having higher order fusion reactions). Most of the stars found today are main sequence stars. Our sun is a typical main sequence star. In the Hertzsprung-Russell Diagram (a plot of luminosity vs. temperature) the main sequence stars stay in the middle of the graph.

White Dwarf

If the mass of a star is in the low or medium category, then it will be a white dwarf someday. Red Giants (described below) after consuming the little amount of hydrogen around their periphery, start collapsing into their core due to gravity. The process is quite fast and a large amount of internal energy is released. Now the star starts its expansion again. But during this process of expansion stars radiate a large amount of energy and get cold. After reaching a certain maximum limit of their size the stars contract for their final fate. Now they do not have any more fuel to burn and due to their small size they cannot start the higher order fusion of elements like carbon. But they have enough energy to look bright. So they are called white dwarfs. The white dwarfs take a long time to radiate their energy and thus to be dark, finally.

The formation of the white dwarfs for the first time was formulated by Subramanian Chandrasekhar. After a number of careful observations and mathematical derivations he found that there is a critical mass for the star to be a white dwarf in its later life. The critical mass after a much accurate calculation was found to be around 1.4 times the solar mass. If a star's mass exceeds this limit then the star will not be a white dwarf rather it will be something else like neutron star or black hole. This critical mass limit is known as Chandrasekhar limit.

Yellow Dwarf

Yellow dwarfs are the small main sequence stars with a medium surface temperature. There are many yellow dwarfs in our galaxy. Our sun is a yellow dwarf. The lives of the yellow dwarf stars are generally stable. The rate of hydrogen fusion is not very fast like the giant stars. In the final stage of their lives they tend to be a red giant.

Red Dwarf

So far according to the general statistics 80 percent or more, of the stars of the universe are red dwarfs. These are small in size and cool in temperature. Due to their small size, their brightness is also quite negligible in comparison to big stars. They generally belong to the K or M spectral types. These stars remain in their main sequence for a long time and can never initiate the fusion of helium. The life span of red dwarfs is the highest among the stars and could be as long as a trillion years (it actually varies from tens of billions to a trillion or more). Proxima is a red dwarf.

Dark Dwarf

The white dwarfs in the process of cooling lose energy and brightness. Finally, they become lack lustre. These lack lustre bodies are called dark dwarfs. They do not have enough mass to start the fusion so they remain dormant till their mass exceeds the critical mass to be able to start another fusion. Many white and dark dwarfs are swallowed by their companions (like black holes and neutron stars) sometimes.

Brown Dwarf

Brown dwarfs are the objects bigger than Jupiter, but smaller than a star (and thus not a star). Due to the insufficient mass they cannot ignite the nuclear fusion. So they are the unlucky objects of the universe until they reach the critical limit for the start of fusion reaction. Sometimes brown dwarfs are found as the companions of stars. Some of the brown dwarfs are believed to have a fusion process at their core at a small scale. If the brown dwarfs are in a surrounding of a planetary nebula, then they have a good chance of promotion to a star by attracting the extra mass from the nebula.

Red Giant

The stars like our sun and a bit bigger or smaller than sun, generally go through the Red Giant stage. When the star's mass exceeds the critical limit to start the nuclear fusion it begins the consumption of hydrogen and convert it to helium. Due to this, the stars tend to expand out ward by virtue of the increasing pressure. So there arises a good balance between the gravity and the pressure due to the fusion process which determines the size of the star. From the size of the star its rate of consumption of fuel is determined. After burning all the fuel the star starts to contract due to the intense gravity of the core. In this process of contraction the star releases some amount of energy which makes the star able to burn the little amount of hydrogen out side its periphery. This process makes the star to expand again and looks quite red. Now the size is larger than the previous size and thus the star is called a red giant. When our sun will be a red giant it will shallow Mercury completely. The stars generally stay in the red giant stage for a smaller period in comparison to their active period. Our sun will spend around 1 billion years in this stage while its active period is around 10 billion years.

Blue Giant

As the name indicates the blue giant stars are very large in size and look blue to us from a large distance. The reason why they look blue is, their surface temperature is too high. In general their surface temperature exceeds 30, 000 K (of course some blue giants are found at a less surface temperature than that). These stars do not have hydrogen lines as the hydrogen atoms are not found on its surface due to very high temperature (all are ionised into protons). These stars perhaps, have consumed all their hydrogen fuel and in their second phase of fusion reaction, they burn their helium storage. So the energy released is more than the fusion of the hydrogen nuclei. After consuming the helium these stars may start any new higher order fusion of some heavier nuclei (but it is found quite rarely). The radii of these stars are quite large in comparison to our sun. But they have a small life time left from being a black hole or neutron star. These stars have a great luminosity, which can be millions of times larger than the sun in some cases. Before being a black hole they may explode as a supernova or hypernova if very large in size and mass. There are many blue giants in our galaxy, but in comparison to other stars their number is quite small. The nearest blue giant to us is the Rigel in the Orion constellation.

Supergiant

These are the largest stars in size and mass. They could appear in various colours (red and blue supergiants are generally found) depending on their mass and age. The spectral type of these stars fall in either O or B category. These stars burn their fuel very fast and thus appear to be very bright. They have also the tendency for going into higher order fusion processes like that of helium or post-helium fusion reactions. That's why they are very hot. Betelgeuse is a good example of supergiant stars (actually now a red supergiant). Its luminosity is more than 10,000 times of the solar luminosity and its radius could be more than 370 times of the solar radius. It is found in the Orion constellation at a distance of 310 light years from earth and it is the main star of the winter triangle (the triangle with Betelgeuse, Sirius and Procyon as the vertices). Supergiants generally end their lives as supernovae and could be black holes at the end.

Cepheid Variable

As we have seen at the beginning of this chapter that the Cepheid variables are very important for the determination of distances and locations of stars in the sky, they really play very fundamental roles in many scientific researches. Cepheid variables are generally, giant and supergiant stars, which in the middle and final stages of their lives become quite unstable. Their size varies due to some internal imbalances, and sometimes oscillates about a mean size. That's why they look as a pulsating star and can attract our attention from a long distance (even from distant galaxies). During the pulsation their size and brightness show an inverse relation (when size increases, the brightness decrease and vice versa). Their pulsation period varies from less than a day to several days. They are the prime distance indicators of the universe. Polaris and Delta Cephei are examples of Cepheid variables.

Supernova

Supernova explosion is one of the most important phenomena of the universe. It is very important to understand the supernovae, in order to understand the formation of neutron stars, black holes, the star lives and their recycle. When a massive star around 5 times of more massive than our sun undergoes the fusion process for years and spent all its fuel it starts shrinking. The shirking is mainly due to gravity and a big amount of energy is released in this process. It enables the star to restart the fusion process with the residual amount of fuel (i.e., hydrogen). Now the star expands a lot to be a super red giant. But as the fuel amount is not enough to keep the star in that state for a long time, gravity starts to take over and the inner core start to shrink. This process makes the star hotter and more energetic due to the fusion of larger elements. Sometimes fission also takes place after the formation of heavier elements like Nickel. This energy is more than enough to combat gravity. A big part of this energy is also transferred to the surface. With this tremendous amount of energy the star explodes like a big spectacular red ball. The energy for this explosion is believed to be coming from the repulsion of the positively charged nuclei. But the explosion is so strong that, it can brighten the whole galaxy (if the star is very big) or nebula (like one in the large Magellanic cloud in 1987) in which it occurs. The masses of the outer layers of the star leave the star for ever after the explosion. These are known as supernova remnants. But they are powerful enough to radiate in the high energy ranges for millions of years. The central part however remains as a highly dense core of neutrons. It is called a neutron star. If the star is too big (15 times or so, bigger than our sun) then it may be a black hole. There is a second way in which the supernova explosion can occur. If a white dwarf absorbs the materials from its companion star then it can be massive enough to explode like a supernova. After each supernova a planetary nebula is born, in which new stars are born by collecting the gas and dust.

Hypernova

Hypernova is similar to the supernovae in some way. But these are quite bigger than the supernova in both its cause and effect. Very big stars generally under go the hypernova explosion. There are many reasons behind the formation of the supper massive stars. Due to the attraction of a big star in its later stage of life its planets and companion stars approach the star as the angular momentum of the star is affected due to the change in size. This process continues till the star explodes itself (as a supernova or hypernova). In some cases the star clusters are controlled by the biggest star of the cluster, if its (the biggest star's) mass is significantly high. Then at some stage the star starts swallowing its neighbours. If this process continues the star can be very large and its final explosion can be a hypernova. It is much brighter and powerful than the supernova explosion. The amount of radiation they emit is quite high and it dominates the rest of the radiations of the universe during that time. Supernovae and hypernovae are thought to be behind the gamma-ray bursts. Generally it is predicted that the stars, which burst as hypernova are more than 20 times larger than our sun. The hypernova radiations are so strong that they have a big impact on the objects at large distances too. Some scientists say that the extinction of the Dinosaurs can be due to a hypernova explosion in some of our neighbouring galaxies. Those dangerous radiations might have made the earth quite unsuitable for the dinosaurs.

Neutron Star

Neutron stars are the end stage of stars, which are generally more than twice as massive as our sun. At the end when the star's gravity pulls everything inward, the outer layers of the star collapse into the core. The process is vigorous due the huge mass and gravity of the big star. The process is so fast and energetic that all the electrons and protons combine with each other to form neutrons. So the star contains only neutrons and thus known as neutron star. In many cases it has been found that the neutron stars are the result of the supernova explosions. In Neutron stars the whole mass of the star is confined in a volume of a sphere having radius around 10 km or so. From this, you can imagine how dense it is. One cc of matter from a neutron star may be billions of tons on our earth. But the interesting thing is that the mass of the neutron star is different from the baryonic mass. It has been found from their gravitational red shift that the baryonic mass is reduced by 20% in the neutron star (thus it is different from Byronic mass; perhaps hyperon-rich matter is there). Neutron stars rotate at very high speed. It can exceed 38,000 rpm. The masses of neutron stars have been found to be around 1.4 times that of our sun. The density can be as high as 10^{14} g/cc. The magnetic field is also quite high (around 10^{12} Gauss).

Pulsar

Pulsars are nothing but the rotating neutron stars having radius between 10 to 15 km and a very powerful magnetic field. As they rotate about their spin axis, they radiate at very high frequency ranges as well. But the name pulsar came from their pulsating properties. They pulse because their magnetic polar axis makes some angel to the direction of our observation. So the powerful radiations which come out from the poles are observed in certain time intervals. One of the noble qualities of the pulsars is that they pulsate (actually rotate) at very accurate time intervals. So they are one of the most accurate time keepers of the universe. The rotating pulsars are charged bodies and their electromagnetic fields are too strong. These pulsars may attract other stars and start a new life. But in general it has been found that they live their lives like a pulsar for a long time, and then due to the accumulation of mass they convert in to different forms. If they exceed the critical mass barrier (to be black hole) they may become black holes, after that.

Black Hole

Black holes are the big brothers of neutron stars. When the mass of a star exceeds 3 times the mass of the sun the star will become a black whole in its later life. So when the star runs out of fuel and cannot sustain the gravitational pull it collapses into itself. In this case, the scenario is quite different. As the star collapses into itself at some point its volume becomes so small that even light cannot escape it. That means nothing can come out of the black hole (actually some other thing also take place and black hole too found to be losing mass in a different mechanism). In terms of the Einstein's general theory of relativity it is a singularity of space-time. That means time is stagnant in the black holes and the mass can have zero volume. This cannot be thought in the general sense. Black holes through out their lives try to suck everything around them. But there is a no return zone around every black hole, anything that falls within that boundary cannot come out of the black hole (but recent findings say that in a latter time the information can be retrieved from the black holes by themselves). It is known as event horizon.

The approximate life-spans of stars and why they end their lives in various forms depend on their mass, rate of fuel consumption and size etc. Overall, we can say that the lives of stars depend very much on their spectral types (more details can also be found in appendix-C). As we know the "O" type stars burn very fast, they die quickly. There are around 0.00002 % O type stars in our galaxy. Similarly, 0.09% B type, 0.6% A type, 3% F type, 7% G type, 15% K type, 73% M type stars in our Milky Way. The death of common stars can also be quite strange. Of course they do not die as supernovae. Supernovae and hypernovae like spectacular explosions happen for the big stars only. O, B, and A types of stars mainly qualify for this. But it has been found that after the final expansion and explosion of almost all the stars (even for a small scale explosion); a planetary nebula is born from the remnants and residues of the dead star. Then another recombination process starts and it helps in the creation of a new star. So the birth and death of stars are some of the eternal processes of the universe.

Quasar

Of course quasars are not stars rather galaxies. But their brightness and luminosity is perhaps due to some stars or black holes. When the radio telescopes were used for the space exploration astronomers found that there are many sources of radio waves in the space. In many cases they were supernova remnants, star birth regions or distant galaxies. But there were some other sources, which were not visible and seemed to be stellar point object. So in the sixties scientists call these types of sources as "quasi stellar radio sources" and later it got the short name quasar. With the improvement of physics and astronomy scientists were able to know that quasars are not only the sources of radio waves rather they are very rich sources of X-rays and gamma rays. Quasars are quite powerful objects of the universe. The invisible nature of some of the quasars is due to the presence of supermassive black holes at their centres. It has been found by comparing the mass of the quasars with those of stars that the central part of the quasars could be up to trillions of times heavier than a star. So the scientists think that only the presence of black holes can make it so heavy. Still the quasars are quite mysterious. Now let us have a look at their radiation mechanisms. Due to the presence of a huge black hole at the centre of the quasar the surrounding space is quite contracted. So the dust particles and gases rush in to the black hole at a very high speed. They get very hot by this fast motion and radiate almost in all frequencies. The radiations outside the event horizon can escape the black hole and travel in all directions of space. The other mechanism can be due to the acceleration of electrons in the intense magnetic fields of the quasars. We know that electrons radiate when they are in acceleration. More about the quasars can be found in chapter 5 and 6.

Planetary System of Stars

So far there is no solid evidence of any planet beyond our solar system (of course astronomers have seen some smaller companions of some stars). Of course Jupiter like companions of many stars, have been observed. But the scientists are not sure, whether they are faint stars or planets or brown dwarfs. The reason of not finding planets is obvious. The planets are really very faint and lacklustre to be seen from the earth (in fact it is difficult to see the distant stars, so planets can easily remain out of sight). But there must be planets to many stars, due to some fundamental reasons. The first reason is that, when the stars are formed the whole matter is not embedded into the star rather some residual part remains outside the star, which form objects like planets. The second reason is, some stars eject matter, due to some reasons like high rotational speed, internal imbalances etc.

In the recent past, astronomers have discovered some planet like objects. In August 2004 some Neptune-like planets (or 10 to 20 limes larger than earth), have been observed around some dwarf stars. They are believed to be planets as brown dwarfs or faint stars cannot be that small. As the planet search programs are going on successfully we should expect more planets in the future.

If we take an approximate number of two planets to each star then, there are trillion trillions of planets in the known universe (within the astronomical reach). The number may be even more. According to the recent predictions, there are around 10^{22} stars in our visible limit. This is actually an average number (100 billion stars per galaxy and thus 100 billion galaxies have 100 billion×100billion stars [more can be found in the next chapters when we shall go through the galaxies]). From these figures you can predict how many planets may be there in the universe.

Life in Other Star System

So far there is no solid evidence of life or life supporting environments in other star system. But we cannot rule out the presence of life in other star systems due to many potential reasons. The existence of life in other star system just needs some similar environments like that of our earth. As our knowledge about the outer solar life is very limited, we do not know what other kind of lives may be possible in other star systems (because our idea is limited to carbon based life systems only). The way we think the intelligent life was evolved in our system, it is not a big thing to exist in other systems. There are many speculations and incidences of aliens in different parts of our world. Who knows, it may have some link with some intelligent lives of other systems. But as our knowledge is limited, we should not speculate a lot. If temperature range and brightness are the main criteria for the existence of life then only G and K stars (see appendix-C for spectral types) qualify for a biosphere. Some F stars may qualify but as the changes are quite fast, they are not the potential candidates for having a biosphere, which in contrast needs a stable condition.

3

Our Home Galaxy

In the previous chapters we have seen many interesting facts of the stars and their companions. But the stars are not the individual objects of the space; rather they are the part of some unified larger collection. Those larger collections are of different types. They can be star clusters, star groups, constellations or galaxies. Out of these the galaxy is the actual scientific term that represents the whole collection of stars because it is bound to a single central gravitational source (that centre is called galactic centre). In this chapter we are going to see our place in our galaxy.

The concept of galaxies first came to the Greeks, Chinese and Indians in the past. They thought that the clusters of stars form a huge collection that is managed by one of the biggest members of the clusters. In fact they did not have the idea of gravitational attraction rather it was thought to be a cause of divine powers. Anyway we find similar mentions by ancient Greek Philosophers and Indian Pundits. But now we can understand many properties of the galaxies, using the physics of large scale. The galaxies can be assumed as a small universe.

You know that our sun is one of the stars of our home galaxy, the Milky Way. The Milky Way is a spiral galaxy, with type Sbc and it is centred with Sagittarius Constellation. As far as the classification of galaxies is concerned, there are mainly three types (types of galaxies can be found in detail in the next chapter) of galaxies. They are spiral, elliptical and irregular. As per the recent observations Milky Way has more than 400 billion stars. Milky Way is one of the galaxies of the Local Group, having 3 large and 30 small galaxies (recent discovery says 37). An amazing thing about the galaxies is that, they form clusters and galaxy clusters form superclusters. There are millions of such clusters and superclusters in the universe (known today). Our Local Group is one such cluster.

As far as the size of Milky Way is concerned it is a huge spiral disc having diameter around 100, 000 light years. Of course this is the approximate diameter of the galaxy; in fact it is elongated even more than that boundary. The thickness of the Milky Way disc is around 7000 light years. Scientist had the idea that Milky Way is a circular spiral galaxy. But recent revelations say that it is not only a circular spiral; rather a bared circular spiral galaxy. That means there exist, some bar like straight structures within the galaxy, which connect the outer spirals with the nucleus or the centre of the galaxy. Our sun is at the outer half of, one of the barred spirals and quite far from the galactic centre. The galactic centre is in the direction of the Sagittarius constellation.

Spiral Arms of Milky Way: We know that Milky Way has a bared spiral structure. It can be understood easily why it is spiral. Generally, old and big galaxies will have spiral structures. We know that when a structure is too big and controlled by a central source, differential velocity will be a common phenomenon in it. That exactly happens with Milky Way. Milky Way has so many stars and huge globular clusters revolving around the galactic centre. So the angular speed of each and every star and cluster and other galactic bodies around the galactic centre cannot be the same. They generally revolve at different angular speeds which give rise to spiral structure. The part of the galactic structure which revolves at same angular speed seems be a non-varying structure, when seen from a long distance. It is known as an arm or spiral arm of the galaxy. There are many such arms of Milky Way. The main arms are

- Norma Arm or Cygnus Arm
- Scutum-Crux arm
- Sagittarius Arm or Sagittarius–Carina Arm
- Local Arm or Orion Arm
- Perseus Arm
- Outer Arm

Actually the arms have been named after the constellations, which contain those arms as their part. Galactic centre is probably at the end of Sagittarius arm and Scutum–Crux arm. Norma Arm is also quite nearer to the galactic centre. The galactic centre areas and the brightest part of the Norma arm are not clearly visible to us due to our odd location. Perseus Arm is just outside the Orion Arm (actually all the arms merge at the centre of the Milky Way) and covers a long area in the outer boundary of the Milky Way. Cygnus Arm is after the Perseus Arm, and most of its parts are away from the galactic centre than other arms of the galaxy. The distances between the spiral arms show a unique logarithmic relationship.

We are in the outer half of the Orion arm or the local arm. Sun is a lone star in the outer half which takes around 250 million years to orbit once around the galactic centre. Our location is in such a part of the Orion Arm that we cannot see the inner parts of the galaxy. Anyway, the surroundings of sun in the spiral arm is quite suitable for us. It is not very dense (in terms of star density) in comparison to other areas of the galaxy. There is no powerful star cluster in the near vicinity and the space is comparatively empty.

The mass of Milky Way is not known exactly. But it is somewhere between 750 and 1000 billion solar masses. Besides stars and star clusters, it has huge mass of interstellar gas and dust particles. The mass or the stars are thought to be less than one third the whole mass of the galaxy. Adding the approximate mass of interstellar gas and dust particles there remains a big unknown mass. It is perhaps dark matter. The stars, interstellar gas and dust generally categorised into three components. They are 1) the halo, 2) the nuclear bulge and the galactic centre and 3) the disc, which contains all the stars and virtually all the gas and dust. The halo consists of very big stars (perhaps very old as well) in very big globular clusters (there are 146 such clusters), which are perhaps the earliest formations of the galaxy. Their age is predicted to be around 10-15 billion years. The rest part of the halo is filled with very hot gases which are in ionized state and are diffused through out the halo. Due to their high temperature they can radiate in the X-ray and gamma-ray ranges. In addition to these properties the actual masses of halos is not known (perhaps a large part is dark). But perhaps the halos contain most of the mass of the galaxy. The discs are the general spiral arms. Our sun is on one such disc which is more than 25000 light years from the galactic centre. It takes around 250 million years (one galactic year for sun) to revolve once around the galactic centre. There are many such discs in Milky Way's spiral arms. It contains mainly young and evolving stars, gases (mainly hydrogen and a small amount of helium) in both ionised, atomic and molecular forms and dust. Due to interstellar dust we cannot see beyond some thousand light years (visible light cannot penetrate distances beyond this to make all the way to earth observatories). So the help of other means (infrared, radio, X-ray and gammas ray astronomy) are taken to know the things beyond that distance. The details of the galactic centre have been discussed in the next section. In local group our Milky Way is the second largest galaxy after Andromeda. But some scientists say that Milky Way is perhaps the most massive galaxy in the local group (there is a speculation as the exact masses of both Andromeda and Milky Way are unknown).

It is still a mystery that, what is there in the galactic centre, that makes the whole galaxy revolve around it. Some scientists say that the galactic centre is very heavy and thus some heaviest objects like the black holes are there. But still it is not clear. The reason is obvious because we are at one of the ends of a spiral arm and the light from the galactic centre cannot come straight to us. Of course some strong radiations and radio waves can penetrate the dense cloud and dust particle layer and make all the way to our observatories. Now let us see some potential evidences and theories about the galactic centre. From observations and statistic of the central region it has been found that the area near the galactic centre is too crowded in comparison to the rest part of the galaxy. The average separation between the stars in the galactic centre area is 1000 AU, which indicates that within 1pc there are about 2×10^6 stars. So the mass is quite huge and collisions are very frequent. There are evidences of large supernova remnants in the galactic centre area. Some hypernova explosions are possible in the galactic centre (perhaps going on always). These energetic processes make the galactic centre incredibly bright. It is not only observed in our galaxy, Andromeda and other big galaxies also show these types of characteristics. There are many evidences of very strong radiations (especially x-ray and gamma ray) in the galactic centre area. So it favours the supernova and hypernova explosions and makes the presence of black holes even stronger. Quasars are nothing but the big galaxies which radiate in high energy region like this at remote corners of the universe. Due to high energy of their spectra they are visible to us from a distance of billions of light years. The measurement of the mass of stars from their speed around the galactic centre area has revealed some interesting facts. Professor Andrea Ghez of UCLA has measured the speed of around 20 stars around the galactic centre and found that they are revolving at a speed of 1000km/s. This tremendous speed can only be possible around a central object having a mass around 2.5 million times the solar mass. According to Professor Ghez it is perhaps a black hole. His results have been confirmed by some other prominent researchers of the field. The central object of Milky Way which has been named as Sagittarius A*(the name is so because it is in the direction of Sagittarius Constellation). From the X-ray studies of 10 light year area of the innermost part of the galactic centre by Chandra Observatory it has been found that a huge area is surrounded by X-rays. It can only be due to the effects of a black hole. The mass also has been estimated according to the observations and has been found to be around 2 to 3 million times the solar mass. So from all these observations and potential theories we can say that black holes are the most possible candidates to be at the centre of our galaxy.

If we see at the energy of the galactic centre it is very interesting that it is a huge source of energy. However we do not know in what form they are and how they are converted from one form to the other (when the internal states of the inner black hole changes). Using the Einstein's equation we can estimate the approximate amount of energy at the galactic centre. It could be around 3 million times the whole energy of the solar system.

Now let us have a look at the age of our galaxy and how it was formed. Milky Way galaxy as we see it today was not like this before some billion years. After its creation it has under gone a lot of changes and evolutions. It is believed that some of the star clusters of Milky Way are among the oldest objects of the universe. That means they were formed at the same time when universe started to cool down and expand. So the creation of Milky Way was started when rest of the universe started. But it was not so mature and big as we see it today. In the early stage it was just a collection of energy and little amount of mass in the form of gas. At that time its temperature was quite high. The evolution took place as the temperature went down. Due to gravity there was accumulation of gases around massive centres. As the mass increased something different happened. Due to nuclear fusion those bodies become too bright and started burning the fuels inside them. In this way a lot of stars were formed and then they started to form groups. Due to the gravitation they attracted a lot of other celestial bodies. As the age passes they were evolved according to the laws of nature (what we now study as physics). After some billion years they take the shape of a spiral (as we find Milky Way today).

Like the planets have their satellites which orbit around the planets, large galaxies have their own satellite galaxies. Our Milky Way has also its satellite galaxies which are quite smaller in comparison to Milky Way. The Magellanic galaxies or Magenalic Clouds (as they are commonly known) are the examples of satellite galaxies. There are some more dwarf satellite galaxies of Milky Way. It happens due to the gravitational strength of Milky Way and the small galaxies are caught by it. Apart from the satellite galaxies our Milky Way is bound with other galaxies of the "Local Group".

Star Clusters in Our Galaxy

Star clusters are quite common in galaxies. There are many star clusters in our galaxy. When the mutual attractions of stars, attract each other and make a star dense region in the space, a star cluster is formed. Looking at some of the star clusters astronomers say that some of the star clusters are not formed, according to that common principle rather they proposed a new mechanism for it. When the stars are formed in the dense molecular clouds of gases and dust, the gas becomes too hot, but due to the gravitational binding, they exchange their energy with each other. Some of the stars get enough energy and thus fled away from the cluster (because their velocity increases the escape velocity of the clusters). But those stars, which cannot escape, remain in the cluster for ever and, consume the gases and dusts. In young clusters there are generally less stars than old clusters. Again in old clusters all types of stars are found, while young clusters lack this diversity. There are many young clusters in our galaxy (also known as open clusters due to their relatively less compact binding). The binary cluster and the Jewel Box clusters are good examples of young or open clusters. In contrast the globular clusters are the examples of old star clusters. They are more dense, compact and spherically symmetrical. The globular clusters generally have 10,000 to more than a million stars in them and no gas is found in the gaps between the stars (as they already been consumed by the stars). M13, M67, M22, Omega Centauri etc are the main globular clusters of our galaxy. There are more than hundred globular clusters in our galaxy.

Omega Centauri is the biggest (known) star cluster of the Milky Way. Its mass is more than 5 million solar masses. It is also one of the brightest globular clusters (brightest in Milky Way and to naked eye) of the local group. Due to its huge mass there are many speculations about its origin. Some astronomers say that it could be a small galaxy that has merged with Milky Way before some billion years. It is around 16,000 light years from earth in the Centaurus constellation. This cluster is well known from a long past. Ptolemy had included it as a star in his catalogue.

Supernovae in our Galaxy

There may be many supernovae in our galaxy as it is a huge spiral galaxy. But the location of our solar system does not allow us to witness those important events of our galaxy. Kepler's supernova is one of them. It is also active now. It was discovered during Kepler's time in 1604. Kepler's supernova remnant is around 13, 000 light years from earth in the constellation Ophiuchus. It is thought to be the last such exploding object in our galaxy. Of course we know about the supernovae of 1572 (it was discovered by Tycho Brahe in 1572, but its remnant is not clearly visible today), Cassiopeia A(it happened before 300 years, at a distance of around 10,000 light years, but its remnant in the X-ray telescopes is clearly visible and quite informative), Vela (now the supernova is not active, but its remnant as a nebula exists, it is also known as Vela SNR, perhaps happened before

10,000 years at a distance of 6000 light years) etc. There are also some records of the historical supernovae, but due to the lack of elaboration they are not well known and their remnants could not be located.

Nebulae of our Galaxy

Apart from the supernovae and star clusters there are many beautiful nebulae in our galaxy. The crab nebula is one of those spectacular nebulae and it is the first object of the Messier's catalogue. There are some other nebulae in Messier's catalogue. Among others the Orion nebula, Helix nebula, Hen 1357, pencil Nebula, Omega nebula, Egg nebula, V838 Monocerotis, Boomerang nebula, Dumbbell Nebula, Trifid nebula, double bubble etc. are the famous nebulae of Milky Way. Helix nebula is the nearest known nebula of earth (it is at a distance of 450 light years) in the Aquarius constellation. There are many such nebulae in our galaxy and many more will be discovered in the future. These nebulae are main areas of star birth.

Boundary of Milky Way

The definition of the boundary of a galaxy is bit complicated. It is not straightforward to say where the boundary of a big galaxy like Milky Way is. There are many problems to the answer of this question. First of all, Milky Way is quite large and we are somewhere in the middle of the radial distance to the galactic centre from which we cannot see clearly the remote ends due to the interstellar dust. Secondly, Milky Way is a spiral galaxy and there are large gaps between the spiral arms at some places. So perhaps some of the arms (or their parts) have not been discovered yet. Thirdly, the gravitational strength of a galaxy like Milky Way is very high. It can attract many small and dwarf galaxies and nebulae. Milky Way has been found to do so frequently. When the small galaxies merge with Milky Way its size increases. However, astronomers and physicists try to find out the boundary where the rules of Milky Way cease to hold good and a new set of rules start. Some significant results have been found with the help of Green Bank Telescope of NARO (National Radio Astronomy Observatory). Researchers at that facility have found that at some distance of 3 to 4 solar radius (the distance between the sun and galactic centre) there is no sign of high velocity clouds (inside the galaxy a lot of clouds of gas and dust are found whose velocities are quite high). There are also evidences of low velocities beyond that distance (the approximate speed of Milky Way towards the nearest supercluster is 200+km/s). Temperatures have also been found to be different after that. So it may be the boundary of Milky Way. Similarly Australian Astronomers at CSIRO have found some thick belts (thickness is around 6500 light years) of gas somewhere 60,000 light years from the galactic centre. Nothing so significant is visible beyond that thick belt of gas. Actually it is the extension of one of the spiral arms of Milky Way and perhaps the border as well. At present the border of Milky Way is assumed to be some light years (more accurate prediction is within 100 light years) from those spiral arms.

4

Our Neighbouring Galaxies

After knowing the interesting lives of the members of our galaxy you must have more interest now to know about the neighbouring galaxies. The diameter of our Milky Way is around 100,000 light years. So our neighbouring galaxies rule beyond that spherical volume (actually our galaxy is like a flat disc not quite spherical). Before going to the details of our neighbours we should have a look at their naming. The naming of the galaxies at the early stages of the astronomy (in 16th and 17th century) was not at all a problem as the number of galaxies known at that time was very small. But as the space technology improved we were able to know about many galaxies. So far there are almost 1 million named galaxies. It is not a small number and the naming process is also not easy. So how they are named? In the past Charles Messier designed a star catalogue. It is known as Messier's catalogue. Later some of them were found to be galaxies and nebulae (or even star clusters). Anyway, it was a great initiative to keep the known stars and galaxies in a catalogue. In Messier's catalogue the objects are prefixed with the letter M. Andromeda galaxy is M31, Triangulum galaxy is M33 and M1 is the Crab Nebula in Messier's catalogue. The first edition of Messier's catalogue was published in 1774 and, covered the objects from M1 to M45. Then Lacaille's nebulae were added into the list when Messier himself confirmed them. The bigger version was published in 1781, which had 103 (up to M103) objects in total (some say 104). Later it was upgraded (of course not by Messier). It was not sufficient when new objects were discovered in the late 19th century. So a new catalogue with a better scope was needed. J.L.E. Dreyer designed the first NGC (or New General Catalogue) catalogue in 1887. Initially it had only star clusters and nebulae as the members. Later it was made more diverse and general. Subsequently, it was regarded as the bigger extension of Messier's catalogue. Dreyer also made the IC or Indexed catalogue, which was more elaborate. Today the NGC, IC and Messier's catalogues are used widely while referencing the cosmic bodies.

But the number of galaxies in the universe is very large. So the naming of all the galaxies is not possible. Rather an approximate number is mentioned. In the Hubble Deep Field there are billions of galaxies. But it is too difficult to name them. So only a very few important galaxies (important from their brightness and size) or galaxy clusters are named and others are described with reference to those main galaxies (or galaxy clusters). Not only in the deep field, well within 60Mpc radius (within the spherical region with radius 60Mpc with the earth at its centre) there are millions of galaxies and quasars. Only around 400 million of the whole galaxies have been kept in a big database (maintained by some universities like Cambridge and Caltech).

Classification of Galaxies

As per the observations it has been found that there are generally, four types of galaxies in our universe. They may be spiral or elliptical or irregular. But there is also a fourth type, which does not come in the above three categories. They are quite different and scientists say that it is due to their young age. After some million years they will come to one of the main three categories. Spiral galaxies have been found in plenty. They are almost 70% of the whole galaxies, followed by 25% elliptic and around 5% irregularly shaped. The reason behind the elliptic shapes is due to the merge of two spiral galaxies. When two spiral galaxies come very close to each other, due to their gravitational attraction, it happens (an elliptical galaxy is formed). Some young elliptical galaxies are also found (due to the presence of two powerful centres or quite elongated star distribution), but as they age, they tend to be a spiral one. It is predicted that someday in the future, a huge elliptical galaxy may be formed, with the merging of Andromeda and Milky Way. Both these galaxies are spiral; they are quite nearer to each other and approaching at a significant speed to merge after some billion years. The irregular shape is also due to the juvenile state. As the galaxies age they come out of their irregular shape and become more regular.

In the present day astronomy and physics the Hubble classification is the most widely used scheme of classifying galaxies. Of course, the above mentioned classification is the result of Hubble classification scheme only. But the actual Hubble classification is more elaborate and diverse (there are subclasses of each category). According to this classification, the galaxies follow the tuning fork diagram. Besides the Hubble classification, there are some other methods of classification of galaxies according to their mass, age and luminosity etc.

Our Nearest Galaxies

Our nearest galaxy is actually variable!! Do not be surprised. It is really true. Before 30 years it was assumed that the Magellanic Clouds are the nearest galaxies of our mother galaxy Milky Way. But recent findings said that it is not true. There are some other small dwarf galaxies nearer to us than the Magellanic Clouds. At the moment Canis Major Dwarf galaxy is the nearest galaxy of our system with a distance of around 25,000 ly (a satellite of Milky Way). Before this it was thought that Sagittarius Dwarf, which is around 81,000 ly (light year) from us (a satellite of Milky Way) was the nearest galaxy till 1998 (until the discovery of Canis Major Dwarf galaxy). The next nearest are the Magelanic cloud system. Large Magellanic Cloud is around 179,000 light years (a satellite of Milky Way) away, while Small Magellanic Cloud is around 190,000 light years (also a satellite of Milky Way) away from us. After that the small galaxies, which are nearer to the solar system are Ursa Minor Dwarf—205,500 ly, Draco Dwarf—248,000 ly, Sculptor Dwarf—254,000 ly, Sextans Dwarf—257,500 ly, Carina Dwarf—283,500 ly, Fornax Dwarf—427,000 ly. The other nearest galaxies which are not satellite galaxies of Milky Way are Leo II—701,000 ly, Leo I—890,000 ly, Phoenix Dwarf—1,271,000 ly, Barnard's Galaxy (NGC 6822)—1,760,000 ly, NGC185—2,021,000 ly (a satellite of Andromeda), NGC147—2,152,000 ly (a satellite of Andromeda). Andromeda Galaxy (M31) is the largest neighbouring galaxy, which is around 2,363,000 ly from Milky Way. There are a lot of small galaxies around Andromeda, which circle it as satellites. They are M32 (NGS 221)—2,363,500 ly, M110 (NGC 205)—2,363,500 ly, Andromeda I—2,363,500 ly, Andromeda II—2,363,500 ly, Andromeda III—2,363,500 ly. After Andromeda and its satellites, LGS 3—2,477,500 ly (satellite of Triangulum) is the next nearest to us followed by IC 1613—2,494,000 ly, Triangulum Galaxy (M33)—2,592,000 ly, Aquarius Dwarf—2,608,000 ly, Tucana Dwarf—2,836,000 ly, Wolf-Lundmark-Melotte (WLM)—3,064,500 ly. In the next section some details of the big galaxies have been presented. In total there are now 40 well recognised members (there may be even more) in the local group of galaxies. The local group is mainly dominated by the two large galaxies, Andromeda and Milky Way. The local group was first observed (as a group of galaxies) and recognized by Hubble in the twenties of 20th century. All the above mentioned galaxies are the members of the local group. In terms of dimensions the local group is limited within a space of 10 million light years (6 million light years accros).

The Magellanic Clouds

The Magellanic Clouds are the two satellite galaxies of our Milky Way. The Large Magellanic Cloud (or well known as LMC) is around 179,000 light years (some other estimations say it is around 160,000 light years) and the Small Magellanic Cloud (well known as SMC) is around 190,000 light years away from our galactic centre. These two small galaxies are generally visible to the naked eye in the southern hemisphere only. So they were discovered quite late, by Magellan in 1519, while he was going around the world to prove that its shape is round. But it is believed that, it was known to the southerners since a long past. When observed in a telescope, it seems to be a broken part of Milky Way due to its proximity.

By shape and geometry these two are irregular, dwarf galaxies and orbit our Milky Way as satellite galaxies. The irregular structure may be due to two reasons. Perhaps, these two galaxies are quite young and evolving to be an elliptical galaxy in the future. Or their elliptical structures have been ripped by the gravitational effects of the big galaxies of the Local group (mainly Milky Way), of which they are the members. They have many hot nebular complexes, due to the presence of big energetic stars. So they radiate a large amount of high energy radiations in the region of ultraviolet and beyond. It makes the surrounding interstellar gas and dust particles to glow, which is quite spectacular from the earth observatories. There are around 2×10^9 stars in the Large Magellanic Cloud and small Magellanic Cloud has even less than that. The most spectacular event in the Large Magellanic Cloud occurred in 1987 when a large supernova explosion took place. It was the nearest supernova explosion in last 400 years. That explosion is still active in LMC. The Tarantula Nebula is the main part of the LMC. It is also known as NGC 2070 in the Dreyer's catalogue (of course the recent version has been modified from the original version). In the past, people assumed it as a star. But Abbe Lacaille for the first time confirmed that, it is a nebula. This nebula is quite bright as there are many big stars and star clusters.

The supernova explosion of the LMC was due to the fiery end of the giant star Sanduleak -69 202 as a huge out burst. The interesting thing here is about the time. The explosion took place a long ago. Around 169,000 years ago. But the lights of the explosion reached us in 1987. So the name of the supernova is 1987A.

Andromeda Galaxy

It is widely known as the M31 of the Messier catalogue. It is the largest galaxy of the local group and along with Milky Way these two galaxies dominate the local group. It is more than 2 million (actual estimations vary between 2.3 to 3 million) light years away from us and perhaps the most distant object visible to the human eye. In the night it appears as a fuzzy patch of light. The whole length of the galaxy covers almost 3^0 of the sky from earth surface. What we see (by naked eye) from earth surface is just the central brightest spot of the galaxy. Due to its clear visibility, Andromeda is known to the people of various parts of the world since last 2000 years. But the first mention in any book is found from the Persian astronomer Abd-al-Rahman Al-Sufi's book, who described about it in 10th century in his "Book of Fixed Stars" as "little cloud". It has perhaps more number of stars than our Milky Way. In size also it is bigger than Milky Way, with diameter around 125,000 light years.

Andromeda is the most studied galaxy after our Milky Way due to various reasons. It is one of the biggest galaxies; we can see its centre from earth better than we can see ours (due to the interstellar dust in the direction of our galactic centre), there are many common phenomena of the galaxies which can be observed from Andromeda. Supernova remnants, spiral structure, huge nebulae all are found in Andromeda. So the common features of a galaxy can be studied from Andromeda quite easily. Besides these, Andromeda is too special due to some of its own features. It has the brightest cluster of the local group. It is said to have two centres. It has some black holes, which are some of the wonders of the universe.

Andromeda galaxy is moving in the space at around a speed of 300km/s. from observations it has been found that Milky Way and andromeda are interacting with each other and thus their speed is quite dependent on each other. They are approaching each other at a speed of 118km/s (the relative distance is reducing at this approximate rate). That means, after some million years (more accurately it could be more than 2 billion years, according to present estimation) both of them will come in contact and the resulting galaxy will no more be a spiral galaxy; rather most probably, will be an elliptical one. Though andromeda is bigger in size than Milky Way, its mass may be less than Milky Way's. It is due to the high density of Milky Way. The recent observations reveal that the mass of the halos of Milky Way are more massive than that of Andromeda's.

Now let us have a look at the centre of Andromeda, which is a subject of curiosity for the physicists and astronomers. Recent observations by Hubble telescope has discovered that Andromeda has got two centres rather than one. But the separation between these two central massive regions is not big. They are jointly around 30 to 40 light years wide. Close observations say that these two central parts are moving towards each other and after some years may merge with each other. Thus one of the potential reasons behind two centres could have been resulted from the merging of another galaxy with Andromeda, a long ago. The main central core of Andromeda is a huge black hole of around 30 million solar masses. So it is one of the richest sources of X-rays and gamma rays of Andromeda. Some other big radiation sources also have been found in Andromeda.

The brightest globular cluster of the local group is G1, of the Andromeda galaxy, whose distance form earth is around 2.9 million light years. It is even brighter than the brightest cluster of Milky Way (i.e., Omega Centauri). It is also known as Mayall II. It has a lot of (Hubble estimation is more than 300 000) big and luminous stars. It is also one of the oldest star clusters of the universe. Thus astronomers believe that it must have some valuable information of the early universe. Milky Way has some big clusters, which are of the same age as G1, but they are not that big or bright like G1. The brightness could be, due to the presence of a lot of old helium burning stars.

Like our Milky way Andromeda has some satellite galaxies due to its huge mass and gravity. M32 (also known as NGC 221) and M110 (also known as NGC 205) are two such satellite galaxies of andromeda. M32 is affecting Andromeda gravitationally and it is also approaching Andromeda very fast.

Triangulum Galaxy

It is one of the big galaxies of the Local Group (actually the third largest after Andromeda and Milky Way). It is well known as M33 or the object number 33 of the Messier catalogue. Of course it was not discovered by Messier; rather Hodierna discovered it in 1654, almost hundred years before Messier. This is a spiral galaxy smaller than both Andromeda and Milky Way. It is around 3 million light years (Hipparcos satellite corrections) from us. It has around 10 billion stars of various sizes. Some of them are very luminous, suggesting the presence of supernova. Of course some supernova remnants have already been discovered in this galaxy. There are also some energetic sources of radiations in this galaxy. Globular clusters have also been observed in M33, which are as big as the clusters of Milky Way. Its mass is predicted to be somewhere around 40 billion solar masses. But in dimension, it is quite large. Its diameter is around 60 000 light years (which is more than half of Milky Way's). The centre of Triangulum galaxy is also a black hole like that of the other big galaxies (but is not a supermassive one, according to the recent studies), which controls the spiral arms of Triangulum. It has been categorised as a Sc or circular spiral galaxy. The distance of Triangulum is around 750 000 light years from Andromeda. Like Andromeda this galaxy is also approaching Milky Way at around 180km/s (the relative separation is reducing at this rate). So in the future, perhaps all the galaxies of the local group are going to merge.

Like Milky Way and Andromeda, Triangulum has its own local group (with its satellites) at a smaller scale. A small member of the Local Group, LGS 3 is perhaps a satellite of Triangulum. Some more dwarf galaxies may be orbiting this galaxy.

Other Members of the Local Group

Other members of the local group are quite small in size and mass. Most of them are irregular dwarf galaxies. Many of them are the satellites of the bigger galaxies. But there are evidences of globular clusters and spectacular nebulae in those galaxies. Even the Cepheid variables have been discovered in those smaller galaxies. Generally big galaxies of the local groups have black holes at their centres. But in case of some of the small galaxies, perhaps there is no black hole or small black holes at their centre. Some such findings have been confirmed in case of the M33 and some other small galaxies. The centre of M33 or the Triangulum is perhaps a new budding black hole. Similarly, the irregular galaxies lack the presence of black holes at their centres. When the black hole becomes mature the galaxy approaches a regular shape.

Destructive Effects of Our Home Galaxy

Our galaxy Milky Way has been found to be destroying its neighbours. It is really a strange and funny thing. Due to its large size, our galaxy is able to attract a large number of satellite galaxies. Many of them are small and dwarf galaxies. Those dwarf galaxies, which revolve around Milky Way, gradually come closer to it and finally collide with it (due to the strong gravity of Milky Way). Of course the collision is not like the collision of the stars and planets. Rather it means the gas clouds and some parts, come within the specified boundary of Milky Way (in the future the collision may be possible at the star level). Even those galaxies, which are a bit larger and far from the colliding distance of Milky Way, are ripped apart due to the huge gravity of Milky Way. It is due to the two centres of attraction. One is of the small neighbouring galaxy and the other is of our Milky Way. These types of destructive effects have seen in case of the Magellanic Clouds. Some parts of these galaxies are getting apart from them, which is a clear indication of the gravitational effects of Milky Way.

We know that universe is expanding against gravity. But in the local group it seems to be an exception. From observations it has been found that Andromeda and Milky Way the two largest galaxies of the local group (the other small galaxies are generally satellites of either Milky Way or Andromeda) are moving towards each other due to their gravitation (their relative separation is reducing). Sometime after 2 billion years, the two galaxies will merge with each other and they will lose their spiral structure after that merging. The combination of these two galaxies will be a huge elliptical galaxy. It is mainly due to their proximity, which is in the gravitational captive zone (gravity perhaps overcomes the expansion of the universe). The separation between Triangulum is also diminishing vary fast. Other small galaxies are also merging with their primary galaxies (i.e., Andromeda or Milky Way). But as a broad scenario, our local group is moving towards the local supercluster of galaxies, (the Virgo cluster) in the direction of the Hydra constellation. The process of approaching each other in the local group is due to the local supercluster and some other powerful sources beyond that. Our local group moves at a speed of 600km/s towards the Virgo cluster. So as a broad scenario, our local group is separated by a very small distance and it does not violate the expanding universe.

5

Beyond our Neighbours

In the previous chapters we have gone through our home and neighbouring galaxies and now let us have a look at the outer space, which is far away from our home galaxy. Till date the total number of galaxies well known (well recognised and under research) to our scientists and astronomers, is around 1.5 million. The present technology however predicts that it could go up to 6 million in a few years. In our local group we have only 33 galaxies (40 including some smaller ones). Rest of the galaxies are mainly distant ones. In fact we do not know the total number of galaxies.

So far there are more than 100 billion galaxies have been predicted to be well within our visible limit (some of them have been observed by the most powerful telescopes, but all of them have not been named and well recognised) and the number can increase to higher ones with the improvement of science and technology. Thanks to the Hubble telescope, this has contributed the most in discovering the distant galaxies, and revealed many cosmic mysteries.

As we have seen the naming of stars and galaxies in the Messier's catalogue or the NGC catalogue, there are similar naming and cataloguing for the distant objects. Mainly, the galaxy clusters and superclusters are named in a similar fashion. The most popular catalogue of that type is the Abell catalogue. Abell's catalogue has more than 5000 clusters and superclusters of the universe which are quite bright and rich in galaxies and matter. Now some universities maintain databases of the clusters and superclusters in their software systems.

Galaxy Clusters and Superclusters

Galaxies are found in groups and clusters. There are many such groups and clusters already been discovered. Like stars, galaxies too like groups (because of gravity). Our local group is one such good example. In the recent astronomy the well known galaxy clusters are: Virgo, Fornax, Leo, Hydra-I, Centaurus, Perseus, Puppis etc. Of course Coma Berenices, Canes Venatici and Ursa Major etc are also well known clusters in the surrounding areas of the local group. The combination of some of the clusters with Virgo Cluster is known as the local supercluster. But the superclusters become a real power, when their clusters get very close to each other and merge. Our nearest cluster is the Virgo cluster, which is approximately 40 million light years away from earth. The centre of the Virgo cluster lies somewhere 50 million light years from earth. There are two giant elliptical galaxies in the centre of the Virgo cluster. Fornax is the next nearest and brightest cluster (with a distance of 55 million light years). It can be observed to the naked eye.

Galaxy Clusters perhaps, are the largest gravitationally bound object in the universe. They have three major parts within them. First of all, they have hundreds of galaxies (in some cases even thousands) within them, containing stars, gas and dust. Secondly, they have vast clouds of hot gas (whose temperature could be around 30-100 million degree centigrade), which cannot be seen through the optical telescopes. Thirdly, they have a huge amount of dark matter (a mysterious form of matter that shares almost 95% of the universe), which has escaped the direct detection by any telescope till today. But its presence is felt through its gravitational pull on the galaxies and the hot gas clouds. The mass of the hot clouds in the cluster, are predicted to be comparable with the mass of all the stars (or may be more than that) of the clusters. The hot gas covers the whole cluster and fills the space between the galaxies. The heating of the galaxies is due to the formation of the galaxy clusters (which come from a long distance at a high velocity to form the cluster and frequently collide with each other violently). The dark matter may also have some role in such a huge temperature. But from calculations it has been found that the mass of the galaxies and the hot gas clouds are not enough to keep them together. So in order to keep them together some 10 times more mass is required. Perhaps that extra mass is provided from the dark matter.

No one knows how long it takes to form a galaxy cluster. It must be taking a long time (billions of years) to form such huge clusters. The main factors, which are responsible for the formation of the galaxy clusters are the amount of dark matter in the cluster, whether the dark matter is cold or hot, how far the galaxies are going apart (how fast the universe is expanding) etc. But the unknown characteristics of the dark matter are making this even more complex to understand. Astronomers now say that the prediction of the pressure of the hot gases that surround those clusters could reveal the amount and characteristics (mass, density and state) of dark matter in the clusters.

The studies of the galaxy clusters can reveal many wonders of the universe mainly the most important ones like the age and size of the universe. It can also give a good idea about the future and whether the present universe is eternal or will go through some critical time. The feat of the gas mass in the galaxy clusters is the key to understand the processes in the clusters and the role of the dark matter in it. It is even more interesting that the galaxy clusters get larger and larger with the merging with other clusters. One such incidence has been found in the cluster called Abell 754. This cluster is around 800 million light years from our earth and does not lie in the plane of our Milky Way. So it is easily visible to the observers on earth. It is in the process of merging with another cluster. One of the clusters has around 1000 galaxies and the other gas around 300. In the process of merging the surrounding gas cloud has been heated to around 100 million Celsius. It is one of the mega mergers of the present time, which might have begun before 300 million years according to experts of the field.

Now you can imagine how massive the galaxy clusters can be. An average galaxy cluster can be million billion times heavier than the sun. So you can imagine how strong is their gravitational attraction can be. This attraction is also responsible for the high temperature of the surrounding gas clouds. When the temperature rises to millions of degree centigrade the radiations from the clusters are in the high energy region only (i.e., X-ray and gamma ray). Thus it can only be detected by X-ray and gamma ray techniques.

As we have seen in the previous chapter Andromeda galaxy and some other small galaxies of our local group show a little blueshift (see appendix-B). That means they are approaching our galaxy. But what could be the reason behind this strange behaviour, while other galaxies are going away from each other. The answer is not very clear at this moment. Some scientists say that, when the two galaxies are not too far from each other, their gravitation wins over the universal expansion. So instead of going away from each other they start approaching each other. If nothing comes in between them then they generally merge. That is the reason behind the big elliptical galaxies. Andromeda and Milky Way, if merge someday, the resulting galaxy will be a huge elliptical galaxy. So gravity is the main reason why galaxies collide with each other in an expanding universe.

There could be another reason behind the blueshifts of the Andromeda and other local group galaxies. If we see beyond our local group, we can find a lot of clusters of galaxies. Some of them are quite large in comparison to our local group both in number and mass. There could be a big attraction from those big clusters. It has been found that our local group is moving towards a local super cluster (the estimated speed is around 600 km/s). But it is strange that it is not heading towards the supercluster directly rather the effects of the "Great Attractor" pulls the local group quite significantly (it is another big supercluster of galaxies behind the local supercluster). So the dual effect of these two gravitationally powerful centres is affecting the direction and velocity of our local group. Here some scientists say that the blueshift of the local group galaxies could be due to the gravitational effects of the local supercluster and the Great Attractor (because the gravity is not equal on each of the members of the local group). The imbalanced gravitation may be forcing the local group galaxies closer. But this possibility is not accepted by some astronomers as the distances of the local supercluster and the Great Attractor is too large.

Clusters can be regular or irregular according to the distribution of galaxies in them. For example the local Virgo cluster is an irregular cluster as the galaxies are distributed in an irregular fashion (the reason is that the cluster is under formation and after some billion years it will be regular). Irregular clusters can have mass between 10^{12} to 10^{14} solar masses. On the other hand the regular clusters have a concentrated central core and nicely formed spherical structure. The Coma cluster is a good example of the regular clusters. The mass of regular clusters is generally more than 10^{15} solar masses. Their size is also quite large and can be in the range of 1-10 Mpc. Clusters when get matured, attract the smaller clusters and superclusters are resulted.

We have mentioned about the local supercluster in the previous section. It is due to the mutual attraction of the Virgo cluster, and other smaller clusters around it. But the bigger clusters like the Great Attractor are one of the most powerful objects of the universe. Because of the Great attractor many galaxies, group of galaxies and clusters are running towards Great Attractor. It can also be seen as a great river of galaxies.

The supercluster, which is attracting our local group towards it, is the Great Attractor. It is so large and powerful that its attraction overcomes the effects of the local supercluster (even our local supercluster is heading towards the Great Attractor). The mass of the Great Attractor could be around 5×10^{16} solar masses. It is around a distance of 45 Mpc (some other calculations show it could be around 65 Mpc) from us. But its impact is quite significant to a distance more than that. In the future our galaxy could be a contributing member of this huge supercluster. The prime indication of the Great Attractor is that it could be made up of a large amount of dark matter. Because it is not yet quite visible to us (of course it lies in the plane of our galaxy whose dusts and gas obscure our sight in that direction) though objects beyond that are visible. Recently, it has been found that the Norma Cluster lies somewhere near the centre of the Great Attractor, but the mass of the Norma Cluster (a densely packed cluster near the centre of the Great Attractor) is not at all enough to explain the characteristic of the Great Attractor. So it strengthens the presence of dark matter in the universe.

What is the effect of the Great attractor on us? The effects are simple as it has been mentioned above; the powerful gravity is attracting Milky Way along with the Local Group at a very high velocity. The estimation through the Doppler's shift finds this to be around 600km/s with respect to the cosmic microwave background. Our movement is approximately in the direction of the Hydra-Centaurus supercluster towards an unclear massive object, i.e., the Great Attractor.

The Great Wall

The great wall of galaxies is the area where a lot of galaxy clusters and superclusters are found. It is seen in the northern sky and has the brightest galaxy clusters in it. The approximate distance of these clusters can be around 200–550 million light years from earth (some broad estimations say that the galaxies in the Great Wall could stretch 600 million light years across). But the approximate dimension of the Great Wall is very large in comparison with the galaxies, clusters and superclusters (it could be a thin sheet of galaxies with thickness 15 million light years, length around 600 million light years and width around 200 million light years). But the density is not uniform through out the Great Wall area; rather there are large voids in between the clusters. The well known clusters in this area are the Coma cluster and the Leo cluster. Hercules cluster, Corona Borealis cluster and the Abell clusters are also present in the Great Wall region. There are also many other superclusters in the local universe. The Great Wall itself is not a cluster or supercluster rather it is the area in the space that contains so many clusters. In this area the density of galaxies is quite high. Situated at the Centre of the Astrophysics Redshift Survey (a survey that's started in 1977 to study the redshift of galaxies in the universe at a large scale) it represents the regular distribution of galaxies. Previously, it was thought that the distributions of galaxies are not regular unless they are very strongly bound to each other through gravity. But the Great Wall shows some anomaly in this regard. It is one of the big mysteries in understanding the universe and its rules at large scale. Some scientists say that if the whole universe is so regular like the Great Wall then perhaps, we know nothing about the rules of the universe.

Does the Great Wall indicate that the galaxies are going to merge to a larger structure, which is larger than the superclusters? The present theories and observations say "no". There is significant redshift between the clusters and superclusters in the Great wall area. But the strangest thing is that the distribution of the galaxies is not that chaotic as it is expected; rather quite ordered (the actual distribution of galaxies is conical in shape or more exactly it is the combination of two inverted cones joined at their peaks). Here perhaps, we do not have a proper theory to explain it. In the opposite of the Great wall there is also another similar wall like structure of galaxies, called the southern wall. Due to the insufficient data and skills the whole areas of the southern wall and the Great wall have not been mapped completely. There could be other wall-like structures of galaxies in the other parts of the universe. Only after the deep space survey, it will be clear.

Beyond the Great Attractor

Beyond the Great attractor lie some other superclusters. Among them is the little studied Vela supercluster. We do not have enough information about it. Because it is too far from earth and lies behind the Great Attractor (also in the plane of the Milky Way). So the direct light has a very slim chance to reach earth. The light from Vela will definitely be lensed (will bent around due to gravity, for details see appendix-B) by the Great Attractor and the other clusters and galaxies on its path to earth. Deep field surveys are going on at very fast rate. In the next few decades we can reveal more secrets of giant clusters and superclusters in the universe. The recent developments on the deep space survey and the findings have been presented in the last sections of this chapter.

Quasars

In chapter 2 we have seen that quasars are not actually stars; rather very luminous galaxies. The distances of the quasars are also quite large from our solar system. Quasars are quite interesting and mysterious for the astronomers. They are big hopes for the studies of the distant part of the universe. The actual reason behind their luminosity and energy is not known. But it could be due to the presence of the very massive black holes at their centres. These distant galaxies show how the space can be diverse. Quasars are the most potential means of discovering the remote corners of the space. The boundary of the universe is predicted, according to the location of the quasars. They can also give the ideas about the universal laws and their organisation. Of course what we see of those quasars is just their past. We have no idea about their present and future. Here in the coming sections we shall have a look at the properties and features of various quasars.

The Quasar that made history was QSO 3c 273. In the sixties people didn't have enough idea about the extra galactic objects and the most part of the outer galactic area was quite unknown. The scientists who were able to catch the radio waves from the space were very excited. They thought that perhaps, civilised human beings of some other star system are sending those radio waves (similar to the pulsars, pulsars were also unknown at that time). But later it was found to be the stellar objects like quasars and pulsars can easily emit these waves. 3c 273 was the second object to be discovered emitting radio waves. This quasar is found in the Virgo constellation (not in the Virgo cluster). But it is too far from the Virgo cluster and its distance has been estimated to be around 2 billion light years from our earth. 3c 273 is also receding a very high speed (around 48, 000 km/s, which is 16 % of the speed of light). Even at that distance it is one of the bright objects in the sky. Luminosity of 3c 273 has been predicted to be around 2 trillion times that of the sun.

Quasars and the AGN or the Active Galactic Nuclei are closely related. When the centre of a big galaxy becomes very massive black hole it starts consuming the surrounding matter at a very fast rate. These types of centres of the galaxies are known as the AGN. But it has been found that the AGNs are less active than the young black holes. Both the quasars and the AGNs have many common characteristics. Both of them have very powerful black holes at their centers. Both of them radiate at high energy regions, both of them have very hot winds of gas and dust around them, they both can jet particles at speeds near the speed of light.

PKS 2349-014 is one of the powerful quasars seen by the Hubble Space Telescope. It has been found that this quasar is attracting other nearby galaxies and perhaps going to merge with them. All these characteristics of this quasar are really strange. It has even surrounded by some other quasars. Many scientists say it the "smoking gun", due to its extreme power of attraction. Here the question arises that, whether the theory of formation of quasars is correct or will it be amended to a new form? Because the characteristics of PKS 2349-014 is different, from the common quasars, in many ways.

The two quasars, called Q2345+007 A and B, are some 11 billion light-years away and were found nearly two decades ago. They seem to be different, but recently scientists have found that they are not two separate galaxies rather a twin galaxy. It also confirmed that the galaxies and Quasars are quite stable in twins. As it was expected that they could be the parts of big clusters in their back ground, nothing like that were observed in the X-ray telescopes.

The Chandra X-ray Observatory has found some other distant quasars (they are SDSS 1306+0356, SDSS 0836+0054 and SDSS 1030+0524). They are thought to be of the very early period. Their radiation intensities in the X-ray range indicate they have supermassive black holes at their centres. The age of those galaxies, is almost same as that of the universe (they were born in the big bang itself). Further detection says that the black holes are there since the galaxies were just one billion years old. So by this time now, they are (the central black holes) quite well established. The distances of these quasars makes them the most distant quasars (they are at a distance of 13 billion light years) of the universe. The X-ray data also shows that the central black holes are surrounded by very hot gas. Generally the super massive black holes radiate weaker X-rays than the young ones. But in this case it is an exception. The black holes are estimated to be around 1 to 10 billion times heavier than the sun; still they emit very strong X-rays.

The most distant galaxy known today has been named as Abell 1835 IR1916, (the newly discovered galaxy) has a redshift of 10, and is located about 13,230 million light-years away. It is one of the greatest things for the astronomers and Physicists. It is seen at a time when the Universe was merely 470 million years young (according to the current estimate the universe is 13,800 million years old), that is, barely 3 percent of its current age. It can give us lots of information about the early universe. Besides that, it has been known that, this distant galaxy is smaller than our Milky Way and almost near the boundary of our universe. As the light from it, is coming from a long distance we are just seeing its childhood. We do not know its present state. It may be quite different now, than what we see.

Of course it may not be the distant galaxy after some days or months or years as new galaxies are being discovered with the help of improving technology and expertise. The Hubble deep field research and findings indicate that there could be galaxies and quasars well beyond that distance. We shall have a closer look on the Hubble deep field in the next section.

Hubble telescope is doing marvellous job in exploring the galaxies. The distant galaxies are very difficult to detect if their energy (coming as radiations) is not strong enough to be detected. While penetrating though our earth's atmosphere they lose most of their energy. So there was a need to capture them before they are absorbed by the atmosphere. This idea worked successfully, and it is being served quite satisfactorily by the Hubble Space Telescope. The Japanese Subaru telescope is also doing some good jobs in this field. In the Subaru deep field project they have discovered more than 70 distant galaxies so far. European Southern Observatory's Very Large Telescope is also doing quite significant work in the deep field survey. Using this facility, this year (2004) the French scientists of France's National Centre for Scientific Research have discovered the Abell 1835 IR1916 galaxy, which is 13.23 billion light-years from Earth. It is the most distant galaxy as mentioned above. It is really amazing to watch the objects, which are so far from us. A new chapter to the space exploration was added when the X-ray astronomy provided very useful data. There are many sources of X-rays and gamma rays in the universe. They can only be studied through X-ray telescopes. The Chandra X-ray observatory is doing pioneering works in this field.

Deep Space

With the help of the powerful telescopes physicists and astronomers are now able to see the most distant and earliest galaxies. The newly discovered Abell 1835 IR1916 is one such earliest galaxy. The distances of that scale are known as deep space. The deep space search has started just before a few years. Without the search for the deep space we cannot have a proper idea of the universe and its mechanisms. But on the other hand it is extremely challenging to explore the deep space. There are many difficulties in the deep space exploration. First of all we are dependent on the light, which comes form those areas as the source of information. Secondly, that is too difficult to catch those light or radiation as they are extremely weak and contaminated with other cosmic rays. Thirdly, the effect of gravity, red shift variation and attenuation in the space itself becomes the biggest difficulties in their accuracy. So it requires a lot of advanced tools and expertise to catch and analyse them. Our science and technology are in a juvenile state in this direction. According to experts, with the help of Hubble telescope it may take a million years to explore the deep space at the distance scale of the most distant galaxy (i.e., Abell 1835 IR1916).

Truly speaking man is not skilled enough to know about all the galaxies. For example there are a lot of faint galaxies within 200 to 400 million light years, which have been ignored because they are not visible to the present telescopes. The number can be even more than, what scientists are predicting. So there is a long way to go to know about the distant galaxies. As the distance increases the detection becomes more and more complex. Very distant and old galaxies that are now in the form of Quasars are very difficult to be detected as they look like a source of point light source similar to the stars. But actually they are the old galaxies in a different form as we have seen above.

The Life Cycle of Galaxies

In the second chapter we have seen the life cycle of stars. The birth of stars, their evolution and then their demise are some of the eternal facts of the universe. Similar to the stars, do the galaxies have their own birth and death or they are something different? Now we shall see how the galaxies are created, evolved and finally die. Whether they reincarnate or not? If so, then in what form do they do that? It is one of the widely accepted facts that, the universe that we see today was created in the Big Bang. So before big bang, the galaxies were not there. According to the Big Bang concepts they were born only after the Big Bang.

The Big Bang was a unique state of the universe with a special kind of space-time orientation. When it happened, the whole thing that was in the universe was a huge amount of energy (perhaps, confined in a very small volume in some strange form). After the big bang everything started getting cold and then energy was converted into matter. The expansion of the universe started with the Big Bang itself. This expansion terminated the unification of the natural forces for ever and gravity became different form its other companions. Then it was the time for inflation, which made the base perfect for the future universe (we shall go through the inflation theory in the next chapter). The whole amount of energy that was there had to change its state in order to cope with the changing states of the expanding universe. So they were converted into matter from the photons or the energy packets. There was also matter and anti-matter annihilation but as the photons were the dominant physical quantities the revere process was not very effective. Due to some imbalances matter became dominant over the anti-matter (however some physicists think that it is a local case and according to them there are perhaps places in the universe where antimatter has an upper hand over the matter, i.e., the physical bodies like stars and galaxies are made up of anti-matter). In the process of the formation of matter various things were created. Due to the presence of gravity the global bodies interacted with each other and many more things happened to come to this stage. The most significant of them is the attraction of matter towards each other, which helped in the formation of particles and gas clouds. When the accumulation got intense the centre become very powerful (due to gravitation) and it attracted the surrounding gas clouds and particles. As the centres become massive they started producing energy in the nuclear (fusion) reactions. They are called stars. But in a bigger scenario when the stars are quite large and nearer to each other their combined gravitational force is very strong and they can have a great influence on the other surrounding bodies and masses. This is what happened in the early universe and big star clusters got closer to each other. Due to gravitation some of the stars got embedded into the clusters directly and others started revolving the massive centre of mass. In this way the central parts of some star rich areas became very powerful and some times making the central star quite massive. The gravitational impact of this central region was effective to a large distance (initially it is believed to be some light years but now they can be as large as a million light years in some cases as the universe is expanding since the Big Bang). The whole assembly of a large number of stars (many billions) is called a galaxy. Initially when they were formed they didn't have a proper shape. As the time passes they oriented themselves into a stable geometrical structure (like spiral, elliptical etc). More about the formation of galaxies can be studied from computer simulations. One such simulation work at a bigger scale is going on in the Max Plank Society's Computing Centre in Garching, near Munich with the collaboration of many experts of astronomy and physics of Canada, Germany, the UK and the US. They want to simulate the universe at a smaller scale taking 10 billion galaxies in to picture. For this purpose they have designed new algorithms and use one of Germany's most powerful supercomputers, an IBM Unix cluster. There are some problems in doing the whole simulation perfectly and the way the universe is thought to be evolved. Some of the problems have been solved and some of them are compromised with minimum shift in the final effect. Anyway, the simulation work is expected to give some fruitful results in understanding the whole universe.

Galaxies can also be thought as island universes as the space between the galaxies are quite empty and the density of matter in that empty space is negligible. Initially they evolve slowly in isolation and then due to gravity start interacting with the neighbouring galaxies. This lead to the merging of galaxies and they get bigger. Due to the mutual attraction of gravity the gases of each galaxy move at a very fast rate towards the centre of the other. This enhances the star formation rate and opens the road for a bigger central control. If there is already a black hole in the galaxy then the merging gives a good meal to the hungry black hole by feeding a lot of matter into it. This process makes very bright glows for a long time which is visible to a long distance (like in Apr 220 in our local universe).

When the galaxies get mature by attracting a huge amount of matter (in forms of stars, planets and nebulae and gas clouds) they go through a different state. The central influential part becomes more and more massive (sometimes billion times the solar mass or even higher in some cases). That massive star undergoes a different end to be a black hole. Once it is a black hole, the galaxy shows its real power. The sucking process continues and the central black hole gets more and more massive.

Then starts a new era, in the life of the galaxy, which is totally controlled by the central super-massive black hole. The stars from the remote corners of the galaxy and the gas nebulae rush very fast towards the centre (of course in a rotating manner as the angular momentum is conserved). Even sometimes smaller neighbouring and satellite galaxies face the same fate as the stars. Now the galaxy look like a single source of light as the central part is dominant over the rest from every corner. This is now called as a quasar, which is the short form of "quasi stellar radio sources" (we have seen in chapter 2 about their naming). Of course most of the quasars known today are not radio sources rather they emit in all ranges of the radiation spectra (the name is so because they were discovered as radio sources). Many of them are very rich sources of X-ray and Gamma rays. The mechanisms of radiation in case of the quasars are different form the stars and it is mainly due to the black hole effects. The life of the quasar depends on various factors. As it is controlled by the central black hole it is a matter of the properties of the black hole. The decay rate of the black hole and the black hole thermodynamics are the main properties that determine how long the black hole will exist. According to this the luminosity of the black hole varies. More than that the surroundings of the quasars play an important role as the central black hole can attract them and make them its part.

So now we have seen that the galaxies start their lives as gas clouds and collections of stars and end as a massive black hole. But it is also true that all the galaxies do not end as a black hole. Rather they lead a different life. It happens in case of the smaller galaxies. Many smaller galaxies do not have a central black hole (as they are not that massive). For example M33 does not have a supermassive black hole in it. So the stars without a central black hole are bit more independent and recycle themselves as stars and the mass extinction as in the case of bigger galaxies is not the common scenario. But as the small galaxies age they too get large in size and gradually develop the characteristics of large galaxies. Some of them get attracted by bigger galaxies and quasars as a satellite and in course of time become their parts. It is quite common in case of Andromeda and Milky Way who have got some satellite galaxies.

It takes a long time for the black holes to decay. According to the common theory of black holes it was believed that, whatever goes into the black holes are lost for good. But recent researches by many experts of this area including Stephen Hawking have found that the information paradox of the black holes is not true. Rather the black holes in their later life open up to give back all the information they have. So in the similar way the matter absorbed by the black holes are revealed, but in a different form.

6

Our Visible Limit and the Boundary of our Universe

We do not know whether there is anything beyond our universe or not? We do not know exactly in what form our universe is in? We do not have a clear idea of the boundary of our universe. But certainly, we can have something from whatever we can see and percept. Why we do not know, where is the boundary of the universe? The answer is quite simple. We cannot see the boundary of our universe. Why we cannot see the boundary of our universe? Because, we do not have the proper tools to see it. So, how do we see or percept our outside world? The answer is easy, by the help of the electromagnetic radiations. We can see or know about the objects, which are radiating and close enough to us. Then, how close is close?

Let us see how close is close for us to see the outside objects. The universe was originated form the big bang sometime between 13 to 15 billion years ago. Then started the expansion of the universe and everything started moving away from each other. Light (not only visible light; rather all the radiation spectra) is thought to be created in the big bang. According to Einstein light moves at constant speed and its speed is same, as before billions of years ago when the universe was created from Big Bang. Let us assume Einstein is completely correct. So light must have travelled a distance equal to its speed times, the time lapsed in between. So, from this simple calculation we can find that light has travelled something of the order of 10^{26} m (in these 13 billion years). So what ever light we are getting is only from that imaginary sphere with our earth at its centre (of course that radius is increasing at a speed of light, as light is moving restlessly). We cannot get light beyond that distance. This indicates, we can see only up to that distance. We do not know, what is there beyond that distance. So we can say that our visible limit or the maximum range, to which we can observe, is limited within that sphere.

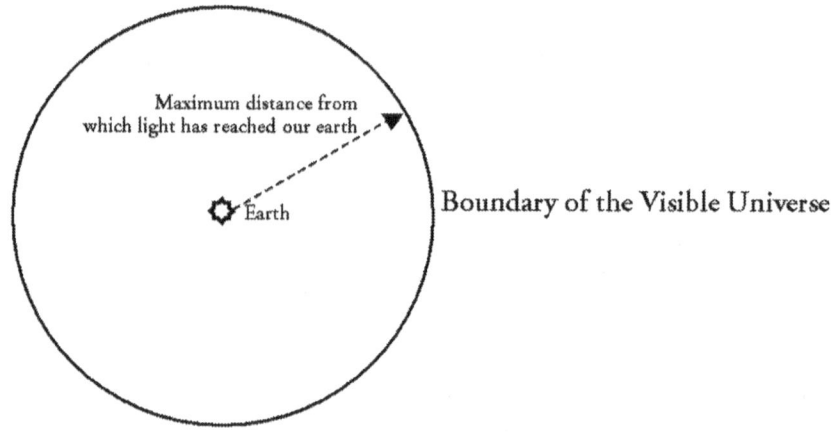

Figure showing the boundary of the visible (or observable) universe, till which we can see

So we cannot know, what is there outside our visible limit. The other problem in seeing the outside universe is the expansion of the universe. Whatever light is coming from the boundary of our visible sphere is not a stagnant source. Everything in the universe moves. So the light from the boundary of the universe gives the information of the past and we do not know its present and can know about it only in the future. Thus it is the natural limit imposed on us in watching the distant objects. Can we go beyond this problem somehow? At this moment looking at our scientific tools it does not seem so.

What could be their beyond that limit? To be honest, it is really impossible to say what is there after the visible range. It could be anything. In the subsequent sections of this chapter and the subsequent chapters we shall have a deeper look at all the possibilities. However, scientists predict that the universe is slightly larger than that (visible sphere) limit. Only experimental and observational evidences can confirm it. There are no such tools or methods to check these facts, whether our universe is just a bit larger than the visible range or much larger than that.

Models of the Universe

At present the prominent scientists and astronomers, predict some of the possible models of our universe. Like planets, stars and galaxies the universe should have its own shape. Of course there are many fundamental differences between the universe and its smaller components like galaxies.

In the early history the model of the universe was not known properly. People had many blind ideas like the geo-centric model. Galileo and Copernicus gave the heliocentric theory. Then in the 17th century Newton's gravitational theory gave the first scientific explanation about the universe at large scale. But later it was found that, there were many flaws, which could not match many of the experimental observations of the large scale universe. It was rectified mainly by Einstein. He gave the four dimensional space-time model, in his general theory of relativity. It was one of the biggest leaps for science.

The present models of the universe are mainly based on the present pictures of the universe and the evidences of the past happenings. Of course the main ideas behind these models are based on some popular physical theories. Out of all theories, Einstein's general theory of relativity plays the biggest role. The fundamental ideas are:

- Whatever we see today, the matter around us, the space-time and almost everything came into existence from almost nothing.

- That perhaps happened some time, before 13 to 15 billion years (some say 15–20 billion years) ago in a big event known as the Big Bang.

- Since after the Big Bang the universe is expanding and the space-time is changing accordingly due to the rearrangement of matter and energy in it.

- The expansion did not happen into void (not happening now too) rather it is a unique swelling or stretching of space-time.

- At the starting the universe was very much hot and unified. All the natural forces were just a single interaction.

- As the cooling process went on, the unification was lost and the ramification of forces came into picture.

- The early universe had no matter at all; rather it was a big bundle of energy in the form of very energetic photons.

- As the cooling process took place, photons were changed into matter according to the mass-energy equivalence principle.

- Since after that matter started forming bigger entities like the nebulae and galaxies as we see today. Gravity played the main role in it.

Someone may ask about the evidences of the above points and their validity. Of course there are enough evidences after all these things. So we shall have a look at the evidences now. The first and the strongest evidence of the Big Bang is that, the universe is expanding. It was first observed by Edwin Hubble and after him almost all have found it to be true. If the universe is expanding then the size of the universe is increasing and in the past it was quite small. In this way if we go backwards in time we shall reach a point, where there is almost nothing. So that nothing is the starting point of the present universe, which was evolved in the so called Big Bang. The second evidence of the evolution of the universe is the conversion of hydrogen into the heavier elements. The principal conversion is the formation of helium from hydrogen in the stars. The present estimation of the amount of Helium in the universe proves that the formation of Helium started sometime around 13 billion years ago. The third evidence is the darkness of the night sky. The darkness of night sky proves that the universe cannot be infinitely old or infinitely large or very uniform (at small scale). Rather it was evolved from some initial state a long ago. The forth evidence is the CMB or the "Cosmic Microwave Background" radiation. The CMB is highly isotropic and homogeneous through out the universe. It proves the validity of the Big Bang theory. It is a sort of black-body radiation and from the calculations it shows that the mean temperature of the universe is around 2.7 K. This shows that the early universe was quite hot and is constantly cooling down.

Based on the above ideas, observations and evidences the following theories have been proposed. Some of them are quite satisfactory and some others are good but not very promising and fail at some special cases. But in fact there is no single theory, which can explain the whole universe from all respects. Anyway, the search is going on and we should hope for the best.

The Inflation Theory

Inflation theory is so far one of the main theories to understand the origin and expansion of the universe. If Big Bang was the real origin of the universe and its constituents, then inflation must have to be there to explain the present universal scenario. According to the inflation theory, just after the big bang there was almost no matter in the universe. All that was called universe was just a huge amount of energy. The nature of space-time was quite strange. All the fundamental forces of the universe were unified. Then started the inflation and a big change took place. Energy was converted to matter and the space-time curvature was determined according to the balance between the matter and energy.

Thus let us see how inflation played the main role in the formation of the present universe and whether inflation is symmetric or not? The Big Bang was nothing but the conversion of some confined amount of energy in a unique fashion. Its space-time warping and other physical properties were something strange. Then something happened (actually what was its nature, we do not know), the whole confined things out burst and started restructuring the space and time. This event is called the Big Bang. We know Big Bang is believed to be the origin of the present scenario due to many potential reasons. But when we go back in time according to the present scenario and expansion rate there is some mismatch. It indicates that there must be different rate of expansion just after the Big Bang which slowed down after some time. So "something" really happened, within that "some intervals or moments" changed the whole thing and made road for the present scenario. This something is called "inflation". During the inflation period the whole differences were made. The expansion was tremendously high. The unification that was there after the Big Bang was lost for ever after the inflation.

The inflation theory was first proposed by a Russian scientist, called Starobinski in 1977 in order to solve the flatness (flatness of our universe) problem. But it remained unnoticed until Alan Guth proposed the GUT theory to solve the flatness and horizon problem. The main idea behind this theory was the inflation. He said the universe must have gone through inflation to be in the present form. The satisfactory reasons behind the inflation are the universe is expanding despite the attractive nature of gravity, the average temperature of the universe is uniform through out, and the density of the universe (galaxy density and thus the mass density) is almost uniform through out (at large scale). Again the universe is dominated by false vacuum. So there happened something in the past that made the universe so uniform. According to Allan Guth the boost for expansion was originated from inflation. It started around 10^{-34} seconds after the big bang and lasted for a very small interval and was terminated around 10^{-30} seconds after the big bang. During this time the universe expanded much faster than the rest of the periods (both before and after the big bang). The prediction of the rate of growth is huge. Perhaps the universe became 10^{50} times larger after the inflation, than before the inflation. Further popularity for the inflation theory came when it successfully solved the horizon problem and the flatness problem.

Why Our Universe is Flat?

So far whatever we have seen and discussed one thing is common, i.e., flatness. The solar system and its members are in a plane. Our galaxy the Milky Way is also a planer galaxy (other well known galaxies are also flat). At the Next higher level we see the galaxies also stay and move in a plane. Why the universe is so flat? This is known as the flatness problem. Physicists and astronomers have been searching for a suitable solution since many decades. There are some theories, which can explain the flatness problem quite satisfactorily. We shall have a look at those theories and how they satisfy the flatness.

Before going to the solution of the flatness problem let us have a look at the facts of our universe. The amount of matter in our universe has allowed it to be what we see today. Our universe does not have enough matter; rather most part of the universe is blank. It is like a victory over the gravity. Had there been more matter, gravity would have been more effective to bind the whole universe to bring back the Big Crunch after the Big Bang. But it did not happen rather the expansion of the universe overcame the gravity and the universe is expanding as we see today (since its birth from the so called Big Bang). Due to the expansion, temperature is going down (but it is not that fast, which can bring an ultra cold situation known as Big Chill). In stead our universe is somewhere in the middle of these two extreme (i.e., the Big Crunch and the Big Chill) situations. Here the obvious question comes to our mind that why the universe is so unique (and its behaviours are so special)?

The flatness problem using the inflation theory was solved by Alan Guth. Guth successfully explained why the universe is not collapsing due the gravity rather expanding faster, and why it is so flat. According to his explanations, the universe is very homogeneous (if we divide the whole universe as huge spherical balls of radius 300 million light years then all the main properties of the balls like mass density, galaxy density and light output etc are almost identical). So after the big bang the universe went through a special phase of expansion called "Inflation". During inflation the expansion was so fast that the universe became homogeneous. The expansion was not equal in all dimensions; rather it was very fast in the planner direction (this is the plane of our present universe) and quite slow in other directions. This made the density of our universe so unique (quite nearer to the critical density). But this theory does not rule out the curvature of space-time as described in the general theory of relativity by Einstein. Within our flat universe there lies the curvature of the space-time and the amount of curvature depends on the amount of mass and energy density. So according to the present situation if the universe is isotropic and homogenous there could be three possibilities of its geometry. It could be either open or close or flat. If it is open the mass density is less than the critical mass density (this prime constant is calculated by observing the expansion of the universe through Hubble's constant and the Newton's universal gravitational constant). If the mass density is more than the critical density then the universe will be a closed universe (where the parallel lines will eventually meet each other if extended infinitely). The third case is the mass density equal to the critical density, which makes the universe to be flat. There are many evidences of the flatness of our universe. The current observations say the density of our space is around 95 % of the critical density.

But according to Guth, the density of our universe should be even more nearer to the critical density. In the future with the help of better science and technology we may find it. As the difference (between the observed density and the critical density) is quite small today, it must be negligibly small (much smaller than the present value) just after the Big Bang (because a slight difference can lead to Big Crunch or Big Chill at a very fast rate, but that is not the scenario). So inflation played the main role in keeping the density so uniform and made our universe flat.

The Ekpyrotic Theory

This is a theory, which explains the probable preconditions of the Big Bang. If so many things happen during the inflation period (which is so small), then is the Big Bang the real beginning? There is some doubt, due to more than one potential reason. The main question is, what could be the possible precondition of the Big Bang? The Ekpyrotic theory has some bold things based on physical facts. It says that the Big Bang was not the origin of the universe, rather just a transition of the universe from one form to the other. It also says that the universe perhaps, has no start or end rather it is eternal (actually oscillatory). It sounds quite different from the previous theories. Isn't it?

Now let us have a look at the Ekpyrotic theory. This theory is originally based on the string theory. The meaning of Ekpyrotic means "conflagration" (ekpyrotic is a Greek word). Actually, the name is based on some Greek beliefs who thought that the universe was created from a fire ball that was cooled down and may repeat again and again. The Ekpyrotic theory has also something similar to these recycling properties. It predicts that the universe is not created once and destroyed once; rather it recycles itself several times, and what we call the origin of the universe, is a transition process or may be the collision of dimensions of the space-time.

However the Ekpyrotic theory has some similarities with the big bang theory. According to both the theories the pre-big bang era (or the pre-expansion era) was cold and empty. Then started the expansion era. In both the cases there are difficulties in explaining the discontinuity. According to the Big Bang, it is the beginning of the universe, while Ekpyrotic theory says it is the transition time in which a contracting universe started its expansion again. Mathematically, the biggest difference is in the strength of the dilation fields.

The original ideas of the ekpyrotic theory actually came from the string theory and the superstring theory. According to these theories the number of dimensions of the universe is not four as we see (three space and one time dimension) and realize rather it suggests 11 dimensions(some scientists predict even more dimensions like 25 or 26). However it is still in a developmental phase. There is no direct evidence of the theory. But the scientists say that there should be some pre-bang vestiges in this universe if the big Bang was not the real origin of everything. The discovery of any such evidence will be the strongest witness of the theory. The most predictable sources for those pre-bangian vestiges are the galactic and intergalactic magnetic fields.

The Ekpyrotic theory has its strong logics from different directions and bases. There are ample chances that some transition takes place in the universe in some definite period. That transition could be in the form of contraction and expansion. That transition could be something what we call the Big Bang. We have seen that almost all; from stars to galaxies everything in the universe has some transition. The stars die and the rebirth process starts in a different way. The galaxies also go through something similar. Then why not in the next larger scale it is similar. The whole thing may undergo a dramatic transition. But one possibility could be that, each transition may not be exactly similar (i.e., the first transition is not necessarily same as the subsequent transitions). But in contrast some of the scientists do not accept the Big Bang in the Ekpyrotic model. They say the expansion and cooling are not due to big bang rather some other reasons may be there.

In addition to the above theories, there are some other theories, but they are not that popular like these two theories. After all, we cannot say which theory is perfect and which is not. There was no witness to the events of the past. There is maximum probability that the present theories will be amended in the future with the development of our scientific knowledge and skills. When we can understand more about the universe we can update our ideas. At last it can be said that the present popular theories are the best theories of the present era. They may be amended in the future.

Centre of the Universe

Is there any centre of the universe or in other words is our universe a concentric one? So far we are not sure about it and cannot say exactly whether it is concentric or not? The implications of the present popular theories say that there is no such centre. After the creation or Big Bang the universe is expanding without a halt. Though geometrically there could be a centre, it does not mean that it is the controlling point of the whole universe. It is clear from the isotropy and homogeneity of the universe in the large scale. If there would have been a centre the universe would not have been so homogeneous. But a few scientists think that there is the possibility of a centre which is controlling the whole universe. According to their views it can be only possible when we can study the motion of all galactic superclusters properly.

Where is the Energy for Expansion?

There is no doubt now, that the universe is expanding. Where from, does the energy come for the expansion of the universe? It is really a mystery. No one knows, where from the energy is coming for the expansion of the universe. Again the rate of expansion is increasing (the universe is accelerating) instead of slowing down. So there should be some huge source of energy for this expansion. Physicists say this is happening by some unknown energy known as the dark energy. There could be a good link between the dark matter and the dark energy.

In this chapter, we have just crossed the boundary of our visible limit, and now going to step into a new world, where only knowledge and imagination rule. That means we cannot get the evidences of anything beyond our visible limit with the help our prime tool (i.e., radiation or light). So far there is no alternative by which, we can know the facts of the remote world. So in the coming chapters we shall discuss the well known ideas of the possibilities of the existence of remote universes.

7

Beyond Our Universe

Is our universe everything? Or something is there, which is not of our universe? It is really a very difficult thing to think and understand. The common perception till now on the universe is that it is expanding and expanding into nothing. What is that nothing? Scientists say that what we say nothing is also a part of the universe. Now one can ask that if that nothing is also in the universe then how the universe is expanding. It is really a good question. What we say expansion today is the expansion of the baryonic matter or what we call visible matter. But the universe is not only composed of visible or perceptible matter rather there is a large fraction of invisible matter known as dark matter. Still today, the dark matter is not known physically. Its characteristics are also yet to be known. What is there beyond our visible limit is really mysterious. What the scientists predict is, mainly based on the facts of the visible universe. As the universe is expanding the separation of the bodies from each other is increasing. The major force, which binds the whole universe, is gravity. Something called dark energy is propelling the universe against gravity and the distances between the objects increase. As the separation increases gravity gets weaker. So now the objects are freer from each other and also from the central attraction of the massive centres (however some local phenomena may be exception due to gravity). If this process continues for a long time at some moment the dark energy will defeat gravity and there will be no order in the universe at all (of course there will be a lot more changes than we can imagine). So now the question arises that where this chaotic universe will head? Will it expand for ever or something strange will come in between? If it expands then, till what limit? If everything is universe including the dark matter and dark energy then are they endless or do they have any limit? Common sense says there should be a limit. In the subsequent parts let us see what the present theories say what can be some other potential possibilities.

Multiverse Theory

Is our universe the lone thing in this creation? Some scientists answer no to this question. So what can be the possibilities? Is there any other universe in this creation and our universe is just a member of that? Could be!!! So how is it possible? Let us see how more than one universe is possible.

There are many implications from various directions that our universe cannot be the only one of its type in the whole creation. So what does this mean? It means that there are many such universes in this creation. Our universe is just a small part of this creation. So if it sounds a bit absurd let us call the small universes as pocket universes, which as a whole form the bigger version of the universe (the whole thing of the creation). So how many such small universes are there? In fact no one knows. It could be infinitely high. Our pocket universe, where our galaxy (along with billions of other galaxies) and our solar system are present is just one of them. So how large can these pocket universes be? No one knows as we have no idea of anything after the visible range. What we call Big Bang may be just a change of phase of our pocket universe.

Now let us see what we mean by a pocket universe? As we have seen that the creation could be very large we have to divide the whole universe into many parts. In the universe, where we are staying is a part of space having some rules (probably a limited number of physical rules). According to those rules its boundary, its behaviour and above all its future are determined. When all these rules start to violate in some part of the universe then that cannot be a part of our universe. It is said to be a different universe (and it cannot be possible in our observable limit, as there is no such big violation been found and there is almost no chance of that). We know that our universe is quite blank (amount of matter is quite small in comparison to vacuum). So there may be a big gap between our universe and its neighbours. Are the other universes are quite similar to ours? No one can say this. But there is a big probability that they have more dissimilarities than similarities with our universe. The gravity could be different in nature (who knows, gravity may not exist at all), the nature of charges could be different, the matter and energy could be different. Or in other words the nature of those universes is beyond our knowledge. So how they are similar to our universe? The answer can only be that they too exist like our universe in this creation.

But multiverse theory indicates that there is possibility of a large number of universes in this creation. It is not a story or mad assumption of anyone. Rather there are many implications from some good theories that there could be many universes. One of the potential theories, which indicate such existence, is string theory. More than this, according to the multiverse theory each universe has its birth and death process (or in other words we can say that they undergo various transitions in their lives if they are eternal). They are created from their own Big Bang (which may be quite different from our own big bang). They age at different rate and different manner according to their own set of rules. But there is no direct way to see them due to our own limitations. Anyway, we should expect that in the future when we will be more smart, we may be able to percept the interactions with those universes.

Multiverse theory makes it possible to understand how the parallel universes can exist in the creation (but now it is not enough). It also makes our ideas quite broad. It is an indication that what we have done so far is quite negligible. We have to go a long way in order to understand the complexities of this creation. But at this time nothing more can be expressed about the parallel universes. Everything will be like a speculation without a real proof.

So now someone can ask the validity and consistency of multiverse theory. Is there any evidence in favour of the multiverse theory? Of course there is no evidence, as it has been said above, except some theoretical implications. We cannot have any physical evidence of anything beyond our universe (actually it is even more limited well within our visible limit as we have seen in the last chapter which is a part of our universe). The astronomical hopes also fade when we cannot get any type of radiations. The outer universes are at such a distance (if they exist) that there is no chance of getting any information from them(or though it is possible, it is beyond our knowledge at the moment). In this context we are too limited and our technologies are too weak. Our theories are also too undeveloped to know the properties of those universes if they exist.

8

Beyond Our Imagination

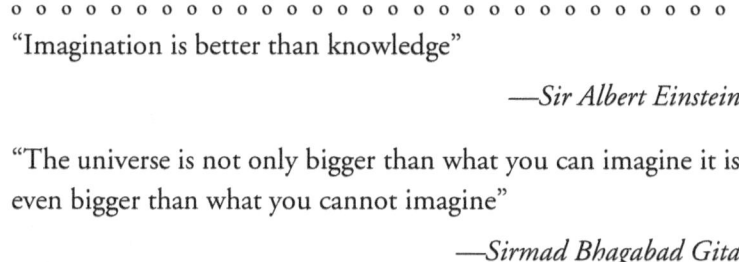

"Imagination is better than knowledge"

—*Sir Albert Einstein*

"The universe is not only bigger than what you can imagine it is even bigger than what you cannot imagine"

—*Sirmad Bhagabad Gita*

"Human Imagination"

—*Mankind's Key to the Universe!!!!*

This is the last section of the book and we have almost looked at the whole thing that we know about the present universe. In the future we will know much more and our understanding will be better. Some new theories may come and some of the old theories may be abolished. So here comes one thing into mind that, is the universe and its surroundings are quite understandable? The question is really very difficult at the moment. After a nice thought, we will say no, to be honest. The reason is obvious, which we shall see later, but now let us have a look at the past of our understandings of the universe. Newton gave his historic theory almost three and half centuries ago. That was a good theory and everyone had good faith on it till the starting of 20th century, when some discrepancies put the physicists in trouble to believe on Newton completely. Then Einstein with his much polished general theory of relativity tried to explain the whole universe with the space-time framework. It explained many miracles, where Newton's theory was a failure. But now many physicists and astronomers do not think that his theory is sufficient to explain the whole universe. They think perhaps there is something more than just space-time.

So why does it happen? The answer is not difficult. Whatever theory is given to explain some physical phenomenon is based on some thorough observations. Does the nature follow the inverse square law of Newton strictly? No. Why? Because nature has a lot more, than just any mathematical expression or theory, based on some postulates. But Newton was not wrong when he put forth his theory. What he observed, he expressed in mathematics and found that everything known at time was following it (approximately). He successfully, calculated the distance, radius and speed of moon. Even today many planets and satellites obey his inverse square law. The satellite technology is based on his theory. But when we make our vision sharp we find that his theory fails. Mercury's motion, the solar layers and almost all the phenomena out side our solar system do not show their support to Newton's law completely. But Einstein found these short comings and tried to sort them out. His special theory of relativity was a great success because it was based on the electrodynamics of Maxwell. Then he tried to bring a good theory to explain the universe. He was successful to a great extent. His theories gave the information about black holes and many cosmological mysteries.

But does the universe obey the theories or mathematical equations strictly. No. Why should it? Universe has its own way of revolution and changes. Whatever theory we give is just our perception based on some good observations. Our mind is not the trend setter of the universe. The universe, to be honest has many more complexities and elegances than what we even cannot think of. What even we observe is just some parts of the trillion trillion.... facts or even less than that. You can realise it better from the following example. Before 2000 years when there were civilisations only at some parts of the world, people mainly believed the geocentric theory (except a few). The reason was straightforward. What they saw, they assumed that to be the truth. But when Galileo proved that it is impossible and heliocentric theory is the actual model of the solar system many people didn't believe it. Then Newton came and mathematically proved it. He opened the vision of people and they got the idea of gravity. Then Einstein found many disparities in Newtonian theory and thus revised it according to the sciences of his time. It was much better than the previous theory of Newton. His idea of space-time and its curvature was also fantastic. So here question arises, why people were not able to understand the space-time concept before 2000 years. The answer is very simple. There was no chance to think of that as their thinking level was limited to geocentric theory only, which they were observing everyday. But we cannot blame them. It was their perception. Newton's ideas were also very accurate at his time. Similarly, Einstein matched his era (the time of 20th century) with par excellence. But do you think that Einstein's theory is the whole explanations of the universe? I do not think so. Einstein's theory is a very nice explanation of the universe and he has really jumped a great leap. But it is up to what Einstein watched and what he understood. Do you think what he imagined was the whole thing in this creation? I do not think so. It is clear from Einstein himself, who told "Imagination is better than knowledge". The reason why I think so, is simple. There is no doubt that Einstein was a great person with extraordinary brilliance. But our imagination has a boundary. What we can imagine is based on our knowledge, experience and our brains' thinking level. Can we imagine about anything, which we have not seen or listened of or our brain cannot even think of? No we simply cannot. Any of our imagination regarding the universe, is mainly based on our visible limit and the present knowledge of the mankind. So our imagination about what we have not seen or known cannot be true and accurate completely.

So there are infinite things beyond our imagination. We have actually got a wonderful tool to understand this universe and our surroundings. What is that? Everyone knows it. It is none other than Mathematics. Mathematics is the most flexible thing in the universe. It can bend like anything, it can expand like anything, it can do miracles, and it can be like anything. What we see in the universe is a kind of mathematical model or other. If the model does not fit perfectly, then it is our lack of skill to frame the perfect mathematics. One more thing is that mathematics is not the brain child of human brain; rather it is the store house of infinite possibilities of the creation.

The Speed of Thought

So far we know that the fastest possible thing in the universe is light. Light can travel restlessly, but it has too many limitations. It follows the space-time curvature; it can be absorbed and emitted several times on its journey; it also interacts like a wave and above all it has a speed limit. So we cannot rely on light or the electromagnetic radiations alone to know the whole universe. In the previous chapter we have seen that we cannot have anything beyond our visible limit. Though the visible limit is spreading at the speed of light it is not enough to understand the whole creation.

So can anything be used, which can substitute light in the process of information gathering? The hope for an appropriate substitute looks grim. There perhaps cannot be any better physical means except light. But the human thought seems to be much more powerful than light. It can travel at any rate. It can be within yourself, it can be within anywhere in the world or universe, it can be sent beyond the universe, and that too within a fraction of a second. Then have we got a solution? The problem is that, it is not a physical quantity. It cannot be sent or gathered like light. But anyway, it is the fastest thing that can transfer information at any rate, at any speed. Thus the main question is now whether we can use our thought in this problem. I think deep on this issue. You too try, who knows, someday we may be successful.

But is the speed of thought infinite? I do not think so. It cannot be infinite. The reason is very straightforward. The thought can travel to the parts of the universe, which are known to us. It cannot be sent to unknown places, which are out of our imagination. So it is the natural limit on the speed of thought. But one thing is sure that it is not a constant; rather a variable.

So from common sense we can say that the theories on universe will change with the course of time. We are too limited to understand the whole universe. Our scientific tools are quite limited to catch the secrets of the universe. Universe is even much more complex than our imagination.

That indicates that there are many wonderful things, which are simply out of our imagination. The presence of other universes cannot be ruled out. What we cannot understand or imagine does not mean that it is not there.

At last as a concluding remark it can be said that we cannot understand everything. We may be very proud of our own achievements, but in fact that is just a negligible fraction of the whole creation. Thus if the human being is really intelligent and smart then, they should realise and admit their limitations. The better solution to these problems is to develop extraordinary imaginations. That can be possible only by big thinking and imagining in all possible ways. Otherwise we cannot understand the universe and beyond that with our available tools.

Appendix A

Parallax and Its Importance in Astronomy

Parallax is very important in order to locate and measure the distance of a cosmic body in the sky. There are many problems in locating a star or planet in the space. The first is due to the motion of our earth. There is nothing in the universe, which is at absolute rest. Our earth is rotating around the sun. So when the position of the earth changes with respect to the sun and other cosmic bodies, we generally do not see other distant objects at their previous places. It is due to the change in their position with respect to the earth (actually it is the earth which changes its position frequently with respect to the distant stars). This problem can easily be solved by parallax. According to this the position of a distant object is calculated according to its position with respect to sun from different locations of earth. Then with respect to some relatively constant stellar bodies the location is noted. This location is comparatively more reliable.

We know in high school, how to calculate the height and distance using trigonometry. In case of parallax also we use a similar calculation. The distance of earth from sun is almost a constant (with a little variation according to the position of earth), called AU or astronomical unit. So any distant star, sun and earth can be the vertices of a triangle. So taking different observations of the star at different times (according to different locations of earth) the distance of the star can be calculated using trigonometry of triangles. The only parameters that need to be calculated for parallax calculation is the parallax angel of the stars (whose distance is need to be calculated). Now there are many good telescopes which can help in calculating the parallax angel to a great accuracy. This method was known to

the astronomers of the past, but due to the non-availability of a good telescope they were not able to calculate the distances accurately. We know
the stars also move significantly with respect to us. But a multiple readings
of parallax can estimate the stars' motion quite smoothly.

The parallax formula is,
D = 1/p
Where D is the distance to the star in parsecs and p is the parallax angel of
the distant star in arcsec (1/3600 of a degree).

Of course parallax methods have their natural limitation. They are not
accurate after a certain distance. It is due to the small parallax angel. As the
distance increases the angel becomes small. So after a few hundred light
years the parallax calculations are not effective. But for the space observatories like the Hubble Space Telescope, it may be up to a couple of thousand light years.

APPENDIX B

Properties of Light

Light was one of the main reasons behind the evolution of biosphere. We have been seeing light since a long past. Still light was not understood properly until the thirties of the last century. Initially light was thought to be made of some particles. But its wave characteristics were quite prominent. So people thought that they are waves. But Einstein made the big break through after explaining the photo-electric effect. Then the wave particle duality was explained by de Brogile and other pioneers of the quantum theory. Since after that light has been one of the important tools for solving many scientific and technical problems. Light has many important properties. As wave it shows interference, diffraction, scattering, polarization etc. Its main physical property is that it is a quantised thing. Light though seems to be continuous in reality it is not. Rather the particle nature is there and it appears in a quantised value known as photon. Different electromagnetic radiations have photons of different energies. For example a visible spectrum of red light has less energy than the blue spectrum. Similarly the photonic properties help a lot to understand the molecular and atomic energy bands. In this appendix we shall see the special properties of light which are helpful and used for the space exploration.

Redshift

Redshift is nothing but the change in the wavelength (and thus frequency) of light in a gravitational field. But according to the formal definition, redshift is the ratio of change in wavelength to the original wavelength. The original wavelength is the rest wavelength. The change is the difference between the observed wavelength and the rest wavelength. When the value is positive we say that it is redshift or the observed wavelength is more than the rest wavelength. That means the energy of the photon has shifted towards a lower value (red light has the highest wavelength and thus lowest energy). If the value is negative it is called negative redshift or blueshift. Depending of the circumstances and surroundings different types of redshifts are found. They are different, because they have different causes behind their creation. So from those differences we can learn a lot about the space from which they are coming. The general theory of relativity for the first time predicted the existence of gravitational redshift. The present formula for the gravitational redshift is also based on the general theory of relativity (because the Newtonian gravitation gives wrong value). If we apply Doppler's principle to the redshift then it indicates that the object that gives redshift is going away from us. Cosmological redshift indicates exactly the same thing, i.e., the universe is expanding. But gravitational redshift is due to the photon's departure out of a gravitational field (i.e., to lower energy). The most interesting thing about the redshift is that, when it occurs in a black hole. In side black holes the wavelengths do not increase rather decrease and cause a blue shift. But we cannot see it as there is no chance of escape from the black hole. But anyway if some one is there inside the black hole somehow he will definitely see the blue or ultra violate or even more energetic radiations inside the black hole. Equivalently, from earth we shall see infinitely large redshifts if some radiation is escaping from just outside the event horizon.

Gravity Lensing

Gravity lensing was first predicted by Einstein in his general theory of relativity. Einstein told that all big objects like stars, galaxies, planets etc warp the space time around them. So when light travels through that path it follows the warps in space-time around those big objects. It was first confirmed in 1918 solar eclipse. Since then it has been used as one of the potential tools for predicting the actual location of the cosmic objects.

Annihilation

Annihilation is the process in which a particle and its anti-particle merge together to result in a photon. Dirac first predicted the existence of anti-particles and Anderson discovered the first anti-particle. That was positron (a particle quite identical to electron except +1 positive charge) or the anti-particle of electron. Soon it was found in experiments that when positron and electron merge, two photons of gamma ray results. That was another verification of Einstein's mass energy equivalence. This property of photons gave some insight about the origin of the universe. The initial universe was perhaps full of photons and matter came out of that.

Appendix C
Classification of Stars

The classification of stars can be done from different basis. For example they can be divided according to their size, mass, temperature, luminosity or some other characteristics. Astronomers and Physicists have classified them in a unique way known as spectral type. The advantage of spectral type classification is that it incoporates all the significant characteristics like mass, size, temperature, brightness and age etc. This is the reason why spectral ytpes are generally mentioned while describing a star. Here we have represented the classification of stars and their characteristics according to their spectral types.

According to this spectral type classification stars are given a designation consisting of a letter and a number according to the nature of their spectral lines which corresponds roughly to surface temperature. But in fact it can represent the overall characteristics of the star in a broad sence. The classes are: O, B, A, F, G, K, and M. O stars are the hottest and the M stars are the coolest. The spectral chacracteristics that determine the type of the stars is mainly dependent on the hyderogen spectra of stars which is mainly due to the excited hydrogen atoms at the surface of the stars(which is directly related to the surface temperature). O types indicate that there is no hydrogen lines and the star is ultra hot (no electron is there with hydrogen neucleus). In case of the B stars the hydrogen lines are detectable but the electrons stay in higher orbits and this indicates that the star is very hot. A stars are also hot but not like O and B types. Their hydrogen lines are detectable and are generally found in the 2nd or higher orbits. "A" stars present very strong and dark hydrogen lines. F and G stars have electrons in the second, first (commonly) and ground levels(very rarely). While K and M stars have their electrons in the ground state and thus they do not present any detectable hydrogen lines. The numbers that are suffiexed with the letters (specifing the spectral types) simply represent the subdivisions of the major classes. Each class is subdivided into ten subclasses (assigning numbers from 0 to 9) depending on their special properties within their class. O and B stars are rare but very bright; M stars are numerous but dim. The Sun is a designated G2 star. According to luminosity ther are eight categories. They are Ia, Ib, II, III, IV, V, VI and VII. Ia represents the very luminous supergiants, Ib represents the less luminous supergiants, II represents the luminous giants, III represents the giants, IV represents the subgiants, V represents the main sequence dwarf stars, VI represents subdwarfs and VII represents the white dwarfs.

In the following table the details of the temperature range of the stars and other properties have been mentioned. Of course all the properties are just the comparaitve notations with respect to our sun.

Spetral type	Brightness	Life Time	Color	Mass	Temperature
O	100,000	5 million	Bluest	20–100	Above 30 000 K
B	500	10 million	Bluish	4–20	Above 15 000 K
A	10	500 million	Bluish-white	2–4	Above 10 000 K
F	2	1 billion	White	1.05–2	Around 7 000 K
G	1	10 billion	Yellow-white	0.8–1.05	Around 5500 K
K	0.2	100 billion	Orange	0.5–0.8	Around 4000 K
M	0.005	1 trillion	Red	<0.5	Around 3000 K

APPENDIX D

Distance Conversion

Here the distance conversion has been represented to have good under-standing on the astronomical distances and sizes of the cosmic bodies. In general the common units of distance like the kilometre or miles are not used to represent the astronomical distances. Because the astronomical distances are very large and sometimes comparative figures are preferred. For extra-solar distances light year and parsec are the preferred units as they are quite suitable for the long distances. Again light is the main information carrier of the distant cosmic objects. So light year representations give a direct significance of many physical things.

1 astronomical unit is the distance between the sun and earth. When the scientists want to provide a comparative figure they generally give it in terms of AU. For example the distance between sun and Jupiter or sun and other planets or two binary stars etc.

AU	Parsec	Light Year	Mile	Metre
1	0.0000048481361	0.000015812507	92955807	149597870000
206264.84	1	3.2615642	19173514000000	3.085678×10^{16}
63241.077	0.30660135	1	5878625400000	9.4607305×10^{15}
1.07578×10^{-8}	5.215528×10^{-14}	1.701078×10^{-13}	1	1609.344
6.6845872×10^{-12}	3.2407788×10^{-17}	1.0570008×10^{-16}	0.00062137119	1

Out of all the above units the most effective, significant and widely used unit is light year. It directly implies the human potential and the distance to know the cosmos.

****

References

References for different chapters have been given in a serial order to make it easy to access according to the chapters. Generally books have been given as references except for a few articles from popular magazines like Scientific American, so that more astronomical penetration is not required.

Chapter 1
Our Sun and its Surrounding

- Nearest Star: The Exciting Science of Our Sun by Leon Golub and Jay M. Pasachoff (Published by Harvard University Press, May 2001) is a very nice text for everyone. It is a good collection of photos, theories and information.

- The Physics of the Planets: Their Origin, Evolution, and Structure by S.K. Runcorn

- Astronomical Tables of the Sun, Moon and Planets by Jean Meeus

- The Solar System (2nd Edition) by John A. Wood of Prentice Hall

- The Solar System Dynamics by Carl D. Murray, Stanley F. Dermott (Cambridge University Press)

- The Adventures of Sojourner The Mission to Mars that Thrilled the World by Susi Trautmann Wunsch

- The Solar System (with InfoTrac and CD-ROM) by Michael A. Seeds

- Solar System Evolution by Stuart Ross Taylor

- Asteroids III (Space Science Series) by W. F. Bottke, Alberto Cellino, Paolo Paolicchi, Richard P. Binzel, William F. Bottke

- A Look at Moons (Out of This World) by Ray Spangenburg, Kit Moser, Diane Moser

- Pluto and Charon, by Malhotra and Williams, edited by Alan Stern and Dave Tholen

- The website of NASA for the facts of planets and sun is very helpful and informative. It is updated in regular intervals. Its URL is: **http://nssdc.gsfc.nasa.gov/planetary/planetfact.html**

- Beyond Pluto: Exploring the Outer Limits of the Solar System by John Davies

Chapter 2
Our Nearest Stars and their Stories

- Stars and Planets by Ian Ridpath (Princeton U. Press)

- The Backyard Astronomer's Guide Second Revised Edition by Terence Dickinson and Alan Dyer

- Observing and Measuring Visual Double Stars by Robert W. Argyle, Bob Argyle

- Red Giants and White Dwarfs; Man's Descent from the Stars by Robert Jastro

- Black Holes, White Dwarfs and Neutron Stars: The Physics of Compact Objects by Stuart L. Shapiro, Saul A. Teukolsky

- The Star Guide: A Unique System for Identifying the Brightest Stars in the Night Sky by Steven Beyer

- Probability 1: Why There Must Be Intelligent Life in the Universe by Amir D. Aczel

- July 2004, Scientific American's article on "The Extraordinary Deaths of Ordinary Stars" by Bruce Balick and Adam Frank

Chapter 3
Our Home Galaxy

- Stars and Galaxies (with InfoTrac and TheSky CD-ROM) by Michael A. Seeds
- The Milky Way As a Galaxy by Ivan R. King
- The Milky Way Galaxy and Statistical Cosmology, 1890-1924 by Erich Robert Paul
- The Milky Way Galaxy by Leonid S. Marochnik, Anatoly A. Such-kov
- Star-Hopping: Your Vista to Viewing the Universe, by Robert Garfinkle (Cambridge University Press, 1994) a good text about the stars of Milky Way.

Chapter 4
Our Neighbouring Galaxies

- Galactic Dynamics by James Binney and Scott Tremaine (Princeton Series in Astrophysics, Princeton University Press, 1987) is an in-depth treatment of the physics of galaxies. Some mathematical and physical background may be required for those who are not familiar with astronomy to go through this book.
- Galactic Astronomy by James Binney and Michael Merrifield (Princeton University Press, 1998) is a good introduction and review especially for the observational properties of galaxies and a must read for the enthusiasts.
- Galaxies and Cosmology by V. M. Canuto
- Galaxies : Structures and Evolution by Roger John Tayler
- The Guide to the Galaxy by Nigel Henbest and Heather Couper (Cambridge University Press, 1994)

Chapter 5
Beyond our Neighbours

- Hubble: The Mirror on the Universe by Robin Kerrod
- Galaxies by Paul W. Hodge is an excellent book on galaxies (Harvard University Press, 1986).
- Beyond the Milky Way: Galaxies, Quasars, and the New Cosmology, by Thornton, Comp. Page
- Voyage to the Great Attractor: Exploring Intergalactic Space by Alan Dressler
- Black Holes Quasars and the Universe by Harry Shipman

Chapter 6
Our Visible Limit and the Boundary of our Universe

- The Big Bang: What It Is, Where It Came From and Why It Works by Karen C. Fox, Karen C. Fox
- Wonders and Their Scientific Explanations by Sudhir K. Routray
- The Inflationary Universe: The Quest for a New Theory of Cosmic Origins by Alan H. Guth
- The Elegant Universe: Superstrings, Hidden Dimensions, and the Quest for the Ultimate Theory by Brian Greene
- The Universe in a Nutshell by Stephen William Hawking
- A Brief History of Time: The Updated and Expanded Tenth Anniversary Edition by Stephen Hawking
- The Road to Reality : A Complete Guide to the Laws of the Universe by Roger Penrose

Chapter 7
Beyond Our Universe

- The Universe and Beyond Third Edition by Terence Dickinson

- The Invisible Universe: Probing the Frontiers of Astrophysics by George B. Field, Eric J. Chaisson

- The Universe and Multiple Reality by Professor M. R. Franks

- The Fabric of Reality: The Science of Parallel Universes—And Its Implications by David Deutsch, David Deutch

- Timeless Reality: Symmetry, Simplicity, and Multiple Universes by Victor J. Stenger

- May 2003, Scientific American's Article on Parallel Universes by Max Tegmark

Chapter 8
Beyond our Imagination

- The Mind of Mankind: Human Imagination-The Source of Mankind's Tremendous Power. By Donald L. Hamilton

COSMIC PHOTO SOURCES

- The great Atlas of Stars Compiled and written by Serge Brunier photography by Akira Fujii

- "The Hubble Atlas of Galaxies" Allan Sandage

- "The Color Atlas of Galaxies" by James D. Wray (Cambridge University Press, 1988) 3-color (UBV) images of 616 galaxies (including all Messier galaxies but M89), taken with telescopes at the McDonald Observatory, Texas, and the Cerro Tololo Interamerican Observatory, Chile, with data and captions.

- "Galaxies" by Timothy Ferris Sierra Club Books, San Francisco, 1980. Superb book (look to get the more expensive full-size edition) with colour and b/w photographs of galaxies and some other objects, from various observatories.

- Firefly Atlas of the Universe by Patrick Moore

- A Photographic Tour of the Universe (Revised Edition) by Gabriele Vanin
- Firefly Guide to Space by Peter Bond
- http://antwrp.gsfc.nasa.gov/apod/archivepix.html (A huge collection of online photos collected by NASA from different observatories and space stations)
- http://grin.hq.nasa.gov/BROWSE/gallaxies_1.html (Photos of galaxies as collected by NASA)

FOR SKY WATCHERS

- A Practical Guide to Viewing the Universe (Third Edition: Revised and Expanded for Use through 2010) by Terence Dickinson

- Splendors of the Universe: A Practical Guide to Photographing the Night Sky by Terence Dickinson and Jack Newton
- Summer Stargazing: A Practical Guide for Recreational Astronomers—For use anywhere in North America through 2010 by Terence Dickinson

OTHER REFERENCES

- QED: The Strange Theory of Light and Matter by R. Feynman
- Faster than Light: The Superluminal Loop Holes of Physics
- Parallax: The Race to Measure the Cosmos by Alan W. Hirshfeld

Index

0-595-33582-9

www.ingramcontent.com/pod-product-compliance
Lightning Source LLC
Chambersburg PA
CBHW030805180526
45163CB00003B/1149